Cambridge Elements ≡

Elements in Emerging Theories and Technologies
in Metamaterials
edited by
Tie Jun Cui
Southeast University, China
John B. Pendry
Imperial College London

A PARTIALLY AUXETIC METAMATERIAL INSPIRED BY THE MALTESE CROSS

Teik-Cheng Lim
Singapore University of Social Sciences

CAMBRIDGE
UNIVERSITY PRESS

Shaftesbury Road, Cambridge CB2 8EA, United Kingdom

One Liberty Plaza, 20th Floor, New York, NY 10006, USA

477 Williamstown Road, Port Melbourne, VIC 3207, Australia

314–321, 3rd Floor, Plot 3, Splendor Forum, Jasola District Centre, New Delhi – 110025, India

103 Penang Road, #05–06/07, Visioncrest Commercial, Singapore 238467

Cambridge University Press is part of Cambridge University Press & Assessment, a department of the University of Cambridge.

We share the University's mission to contribute to society through the pursuit of education, learning and research at the highest international levels of excellence.

www.cambridge.org
Information on this title: www.cambridge.org/9781009246408

DOI: 10.1017/9781009246415

First published 2022

A catalogue record for this publication is available from the British Library.

ISBN 978-1-009-24640-8 Paperback
ISSN 2399-7486 (online)
ISSN 2514-3875 (print)

A Partially Auxetic Metamaterial Inspired by the Maltese Cross

Elements in Emerging Theories and Technologies in Metamaterials

DOI: 10.1017/9781009246415
First published online: August 2022

Teik-Cheng Lim
Singapore University of Social Sciences
Author for correspondence: Teik-Cheng Lim, tclim@suss.edu.sg

Abstract: A partially auxetic metamaterial is introduced, inspired by the Maltese cross. Each unit of this metamaterial consists of a pair of counterrotating equal-armed crosses, which is interconnected to neighboring units via hinge rods and connecting rods. Based on linkage theory, the on-axis Poisson's ratio was established considering a twofold symmetrical mechanism, while the tetrachiral and anti-tetrachiral mechanisms were identified for on-axes uniaxial compression. A shearing mechanism is suggested for pure shearing and diagonal loading of the metamaterial with square array. Results suggest that the approximated infinitesimal models are valid for the Poisson's ratio of the twofold symmetrical mechanism and the tetrachiral and anti-tetrachiral mechanisms under on-axis tension and compression, respectively; however, the finite model is recommended for quantifying the Poisson's ratio under pure shear and off-axis loading. This metamaterial manifests microstructural trinity, in which three different loading modes result in three different groups of deformation mechanisms. Finally, suggestions are put forth for some unsolved predictive problems.

Keywords: Maltese cross, metamaterial, microstructural trinity, partially auxetic system

ISBNs: 9781009246408 (PB), 9781009246415 (OC)
ISSNs: 2399-7486 (online), 2514-3875 (print)

Contents

1 Introduction

1.1 Mechanical Metamaterials

From the Greek word "*meta*" (μετά) and the Latin word "*materia*," the word metamaterial – which is literally translated as "beyond matter or beyond material" – can be defined as any material that is architected to possess desired properties that are not naturally manifested. This definition is sufficiently broad to include electromagnetic metamaterials (e.g. [1–6]), acoustic metamaterials (e.g. [7–9]), and mechanical metamaterials (e.g. [10–12]), to name a few. For more updated information on metamaterials, the reader is referred to some of the recent literature [13–17].

The domain of mechanical metamaterials encompasses structured micro-lattices that exhibit not only extreme mechanical properties, but also reversed mechanical properties. Reversed properties refers to negative or reversed behavior of properties that are conventionally and intuitively positive. Naturally occurring negative mechanical properties are very rare in nature and are not tunable. Categories of negative mechanical properties include – but are not limited to – negative Poisson's ratio materials, negative stiffness systems, negative thermal expansion (NTE) materials, negative compressibility (NC) materials, and negative moisture expansion (NME) materials, which are also known as negative hygroscopic expansion materials or simply negative swelling materials. NTE materials expand and contract upon cooling and heating respectively [18–38], while NC materials expand and contract during an increase and decrease in pressure [39–59], respectively. Analogous to the NTE and NC materials, the NME materials expand and contract when the environmental moisture concentration decreases and increases respectively [60–67]. The previously mentioned metamaterials possess negative values of coefficients of thermal expansion (CTE), compressibilities (or the coefficients of pressure expansion), and coefficients of moisture expansion (CME). These materials respond in the opposite manner in comparison to conventional ones. The negative stiffness system works based on snap-through. For example, during an applied compressive force there is an opposing force exerted by the system; however, beyond a certain threshold the elastic system undergoes snap-through such that the contact point moves away from the loading point. If adhesion exists at the contact point, then a force of opposite sign is registered during snap-through [68–89]. An introduction to, and description of negative Poisson's ratio is furnished in the next subsection.

1.2 Auxetic Materials and Metamaterials

The Poisson's ratio of a solid is defined by the negative of a strain transverse to the load to the strain in the loading direction, that is, if a load of σ_{ii} is applied on the solid parallel to the Ox_i axis, then the Poisson's ratio is written as

$$v_{ij} = -\frac{\varepsilon_{jj}}{\varepsilon_{ii}} \tag{1.1}$$

where ε_{jj} and ε_{ii} are strains along the orthogonal axes Ox_j and Ox_i, respectively. Intuitively and by experience, the application of tensile load lengthens the solids and at the same time contracts their width, and prescription of compressive load shortens the solids and at the same time expands their width. Regardless of whether the applied load is tensile or compressive, the strains of opposing signs would give $\varepsilon_{jj}/\varepsilon_{ii} < 0$. For this reason, a negative sign has been incorporated into the definition described by Eq. (1.1) so as to give a positive value to the Poisson's ratio for (almost) all naturally occurring materials. Suppose the signs of strains in both the loading direction and its transverse direction are the same; then one obtains a negative Poisson's ratio material. The difference between a positive and a negative Poisson's ratio material is pictorially represented in Fig. 1.1.

From the Greek word "auxetos" or "auxetikos" (αὐξητός / αὐξητικός), which means "that may be increased," the word auxetic has been coined by Evans et al. [90]. Perhaps the first allusion to auxeticity (the negativity of Poisson's ratio) was made by Voigt [91] on iron pyrites, but this was later proven to be incorrect [92,93]. Despite this false start, auxeticity was later reported for various solids [94–103]. In addition to the preceding, the negativity of Poisson's ratio has also been measured in biological systems [104–116], as well as in some elements and compounds [117–122], alloys [123–131], and composites [132,133]. The

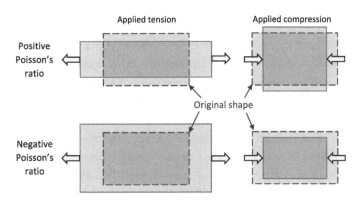

Figure 1.1 Comparison between solids of positive (top) and negative (bottom) Poisson's ratio in response to applied tension (left) and compression (right).

resurgence in research and development of negative Poisson's ratio solids is owed to Lakes [134], who produced foams that convincingly demonstrate the negativity of Poisson's ratio. Fixing other mechanical properties constant, materials with auxetic properties have been shown to exhibit enhanced mechanical properties [135–150] such as greater resistance toward indentation force and shear loads, as well as exhibiting shape memory and improved damping. Discoveries and improvements on various methods for producing auxetic foams and polymers play an important role in permitting further experimental verification on developed models [151–171]. Alongside with the previously mentioned development, efforts have been made to establish a relationship between the deformation mechanisms governing the behavior of auxetic systems from macroscale to molecular-scale [172–176]. The more-salient reports on molecular auxetics include carbon allotropes and related structures [177–187], as well as naturally occurring crystalline structures such as zeolites, silicates, zeolite-type frameworks, frameworks of oxides, and body-centered cubic (BCC) metals [96,97,99,102,188–204]. A number of man-made systems were established based on physical concepts that shed light into the possibility of attaining various types of counterintuitive properties. An example of this is the hard disks models, based on cyclic polymers [205–219].

The following lists the microstructures and/or mechanisms that permit the manifestation of auxetic behavior:

(a) reentrant models,
(b) double arrowhead,
(c) interconnected star models and their related version in the form of interconnected petals (smoothed stars),
(d) nodule-fibril models,
(e) rotating unit models,
(f) slit perforated models,
(g) hard cyclic polymers,
(h) swastika models and other missing rib models,
(i) chiral and anti-chiral models,
(j) interlocking and sliding models,
(k) indented sheet models (which include the egg-rack model, foldable macrostructures, folded or origami sheets, dimpled sheet, and uneven graphene sheets),
(l) helical yarns, plied yarns and stitched-through yarns, auxetic textiles, and auxetic tubular braids,
(m) liquid crystalline polymeric models,
(n) instability-induced auxetic models,

(o) ring-rod assembly models,

(p) linkage mechanism models,

(q) Voronoi structures,

(r) other models such as pre-buckled lattices, star perforations, and composites with hollow tetrahedral inclusions.

Descriptions and elucidations of these microstructures in manifesting auxetic behavior are summarized in a number of reviews on auxetic systems that are more generic [220–232] as well as those that are more specific [12,87,233–239]. Readers are also referred to related monographs that include one on auxetic materials and structures [240], one on auxetic textiles [241], and one on the mechanics of mechanical metamaterials that incorporates auxetic metamaterials [242]. While early investigations on negative Poisson's ratio materials do not exactly exhibit repetitive microstructural patterns, the proposed models were nevertheless idealized such that repetitive unit cells were adopted for analysis. With rapid progress in precision additive manufacturing, it was therefore a matter of time before actual auxetic metamaterials were made to verify the corresponding analytical and computational models.

1.3 Inspired Metamaterials

In some instances, progress in science has been aided by accidental discoveries, and in others by inspiration from biological systems. By way of example for the latter, the skin color of some reptiles is understood to be the consequence of interplay between pigmentary and structural elements [243,244]. Specifically, Teyssier et al. [245] showed that rapid color change in chameleons occurs in part due to active tuning of guanine nanocrystal spacing in a triangular lattice at the upper multilayer. In fact, the skin colors of some animals do not arise from pigments but are instead "structural." If the origin of their color were due to pigments and dyes, there would be no possibility of changing their color, much less in a rapid manner. It is the reorganization of the responsible bio-nanocrystals that permits such rapid alteration of their skin colors. Since the nanocrystal orientation permits a certain range of light wavelengths, and hence color, to be reflected, it follows that a reorganization of these nanocrystals would therefore shift the wavelength of the reflected light, and hence reflect a different color. On the basis that these optical characteristics are structural, it implies that the manifested behavior is more effective if the geometrical properties of the surface structure are well-defined rather than random. This is because the cancellation effect on the physical properties increases with the degree of randomness. Since well-defined geometrical properties are artistic when arranged in a repetitive pattern, it follows that patterns with artistry can therefore provide inspiration for designing metamaterials. Following the works of Teyssier et al. [245], other examples of the design of metamaterials that take advantage of structural or geometrical properties to control the light reflection from surface have been reported (e.g. [246–249]).

Due to their nature, inspirations that give rise to new auxetic and other metamaterials are typically geometrical, including geometrical patterns of a cultural and religious nature. One such category is that of Islamic geometrical patterns, which have been developed from simple designs to complex motifs over many centuries. The earliest attempt to draw inspiration from Islamic motifs to develop new auxetic metamaterials was presented by Rafsanjani and Pasini [250]. In their work, the new auxetic structures were inspired by repetitive geometrical patterns found at the Kharraqan twin tomb towers in western Iran. Concurrently, Rafsanjani and Pasini [251] developed rotating unit models based on square and triangular rotating units with tilted cut motifs, each with straight parallel and circular arc cuts, as furnished in Fig. 1.2. Thereafter, mechanical metamaterials

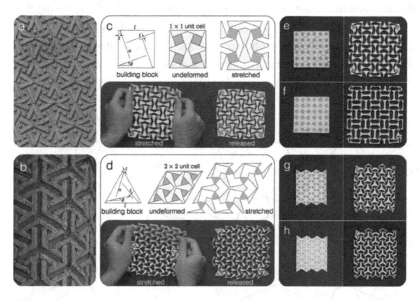

Figure 1.2 Bistable auxetics by Rafsanjani and Pasini [251]: Brick decorations based on (a) square and (b) triangular motifs in the Kharraqan twin tomb towers in western Iran. The building block and the corresponding unit cell in its undeformed and deformed states comprises (c) square and (d) triangular rotating units with tilted cut motifs, which can maintain the stable deformation pattern after the load release. (e)–(h) Undeformed and deformed states of bistable auxetics with square and triangular rotating units for circular and parallel cut motifs. Reprinted with permission from Elsevier.

that were designed by inspiration from Islamic geometrical patterns can be categorized by their deformation mechanisms, such as the sliding mechanism, linkage mechanism, and the curving mechanism.

The sliding mechanism was first proposed by Ravirala et al. [252] based on their interlocking hexagon model. A perfectly auxetic system, in which the Poisson's ratio is $v = -1$, was later developed by inspiration from a type of interconnected eight-pointed star that is commonly found in Islamic motifs [253]. Such an auxetic system consists of two types of rigid stars with slots and sliders, to permit mutual sliding toward and away from each other. See Fig. 1.3.

Many auxetic systems are based on linkage mechanisms, of which the classical ones are known as the reentrant model [134] and the double arrowhead [254]. Linkage mechanisms that give rise to other "negative" properties have also been established. One such example is the modification of the Hoberman circle to produce a 2D NTE structure by Li et al. [255] via incorporating rods of greater CTE, that is, an overall NTE metamaterial is obtained by combining rod elements of positive but highly contrasting magnitudes of CTE. Following this, a type of linkage mechanism consisting of rigid squares, rods of low CTE, and crosses of high CTE was proposed to produce a metamaterial that exhibits conventional behavior during cooling and/or drying, but reverses to NTE and/or NME during heating and/or moistening [256], as shown in Fig. 1.4.

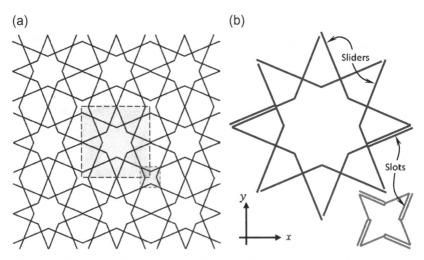

(a)   (b)

Figure 1.3 An Islamic motif consisting of eight-pointed stars in square array interlaced with four-pointed stars (a) from which both types of stars are modified to possess slots (b) so as to permit sliding motion [253]. Reprinted with permission from Institute of Physics.

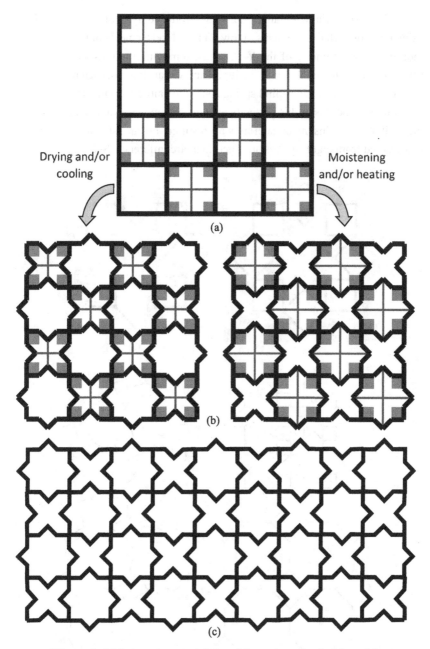

Figure 1.4 (a) A metamaterial that (b) persistently shrinks with environmental changes to resemble (c) an Islamic motif [256]. Reprinted with permission from Elsevier.

A 2D metamaterial with large NTE was introduced by Cabras et al. [257] by using interconnected Y-elements of low CTE and rod elements of high CTE. By modifying the geometry of the Y-elements, removing the rods of high CTE, selecting rigid materials with insignificant expansion coefficients for the Y-elements, and attaching spiral springs onto the paired Y-elements, the effective Young's modulus of this perfectly auxetic system was evaluated [258,259]; the unit cell of this metamaterial was inspired by an Islamic star that is constructed from eight squares arranged in a circumference. See Fig. 1.5.

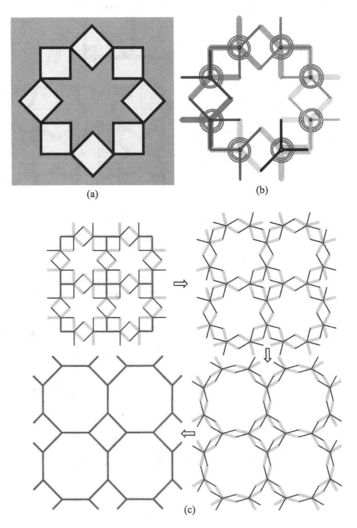

(a)   (b)

(c)

Figure 1.5 (a) An Islamic star constructed from eight squares arranged in a circumference, and (b) a unit cell containing eight pairs of rigid Y-elements. (c) Gradual equi-biaxial deformation upon prescription of uniaxial stress [258,259]. Reprinted with permission from MDPI and Institute of Physics.

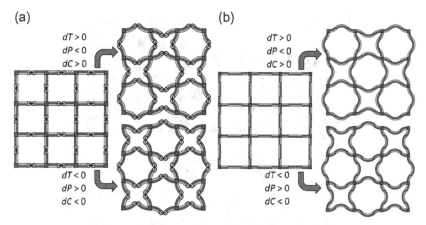

Figure 1.6 Two types of square grids in their original state and their deformed states upon changes to the environmental conditions [262]. The deformed shapes resemble the Islamic motif shown in Fig. 1.4(c). Reprinted with permission from Springer Nature.

The curving mechanism is based on the working principle of the bimaterial strip. Lakes [260,261] was the first to adopt the bimaterial strip for producing metamaterials with tunable positive and negative effective CTEs. The attainment of sign-toggling CTEs, compressibilities, and CMEs that resemble the Islamic motifs in Fig. 1.4(c) can also be approximated using bimaterial strips, as shown in Fig. 1.6 [262]. With reference to Fig. 1.6, the bimaterial strips are made from materials of high expansion coefficients (indicated as brighter strips) and insignificant expansion coefficients (indicated as darker strips).

Drawing inspiration from one of the mosaic patterns found in the Alhambra Palace, as shown in Fig. 1.7(a), a metamaterial with sign-toggling CTE, compressibility, and CME has been designed [263]. As furnished in Fig. 1.7(b), the resulting micro-lattice, arising from environmental changes, produces a hexachiral honeycomb, which is known to exhibit auxetic properties.

Deriving inspiration again from another mosaic pattern found at the Alhambra Palace as well as from a pierced screen of Cordoba's Great Mosque, as shown in Fig. 1.8(a) and (b) respectively, an alternative metamaterial with sign-toggling CTE has been proposed [264]. It is clear from Fig. 1.8(c) that the resulting microstructure due to temperature change gives rise to a honeycomb that resembles the anti-tetrachiral geometry, which is known to produce auxetic behavior.

In addition to mechanical metamaterials inspired by Islamic geometrical patterns, there have been a number of advances made in metamaterials inspired

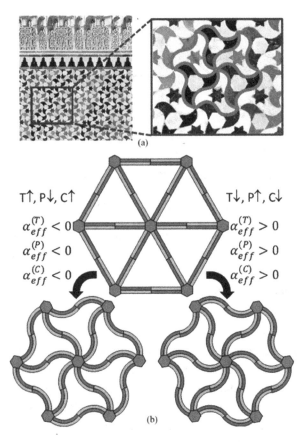

(a)

$$\alpha_{eff}^{(T)} < 0$$

T↑, P↓, C↑

$$\alpha_{eff}^{(T)} < 0$$
$$\alpha_{eff}^{(P)} < 0$$
$$\alpha_{eff}^{(C)} < 0$$

T↓, P↑, C↓

$$\alpha_{eff}^{(T)} > 0$$
$$\alpha_{eff}^{(P)} > 0$$
$$\alpha_{eff}^{(C)} > 0$$

(b)

Figure 1.7 (a) Identification of the basic pattern in one of the mosaic patterns found in the Alhambra Palace, and (b) a unit cell of the metamaterial that shrinks to form this basic pattern [263]. Reprinted with permission from Springer Nature.

by repetitive patterns of various cross symbols, such as the simple equal-armed cross aperture [265]. Most optical/electromagnetic metamaterials that contain repetitive patterns of cultural/religious crosses are inspired by the Jerusalem cross [266–275]. The Jerusalem cross, alternatively known as the Fivefold Cross and the Cross-and-Crosslets, was adopted as the emblem of the Kingdom of Jerusalem from the late thirteenth century.

To a lesser extent, some optical/electromagnetic metamaterials have been designed by inspiration from the Celtic cross [276]. A mechanical metamaterial that exhibits auxetic, NTE, NC, and NME properties has been developed by inspiration from the Celtic cross [32]. Figure 1.9(a) shows a unit of this mechanical metamaterial consisting of a ring and two long rods attached to the inner

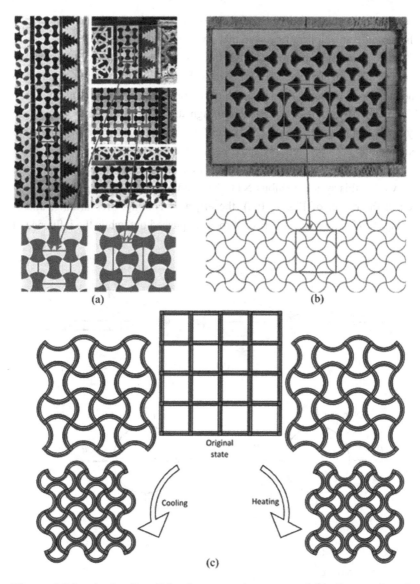

Figure 1.8 Inspiration from Islamic geometric patterns at the tile mosaics of Alhambra Palace (a) and a pierced screen of Cordoba's Great Mosque (b), with the composite metamaterial undergoing intermediate and complete cooling and heating contractions (c) that mimic the Islamic geometric patterns in (a) and (b) [264]. Reprinted with permission from Elsevier.

surfaces of the ring and protruding diametrically on the opposite side. If the rods are pushed as shown by the horizontal arrows in Fig. 1.9(b), then the circular ring changes into an elliptical ring such that it narrows in the other direction. If the rods are pulled as shown by the horizontal arrows in Fig. 1.9(c), then the circular ring changes into an elliptical ring such that it widens in the other direction.

If two pairs of rods are attached, the sign of the Poisson's ratio depends on how the rods are attached to the rings. When all four rods are attached to the outside of the ring, then one obtains a non-auxetic behavior, as shown in Fig. 1.9(d). If a pair of rods pass through the ring, one obtains an auxetic system, as indicated in Fig. 1.9(e) and (f). If all rods pass through the ring, one obtains a non-auxetic system, as denoted in Fig. 1.9(g).

Although the configuration shown in Fig. 1.9(g) is non-auxetic, it can be shown that this system exhibits NTE, NC, and/or NME. Starting from the unit cell at the center of Fig. 1.10(a), the effect of cooling, pressurizing, and/or drying shrinks the ring such that the rods are pushed out, while the influence of

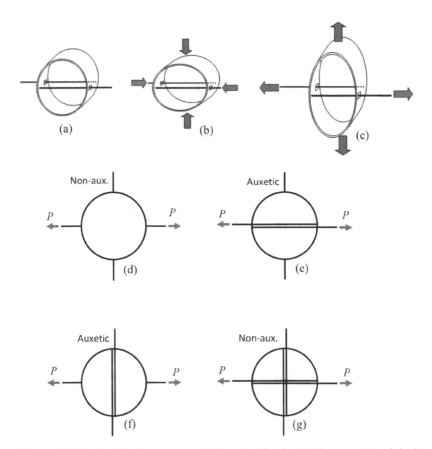

Figure 1.9 A unit cell of a metamaterial inspired by the Celtic cross: (a) original state, (b) under uniaxial compression, and (c) under uniaxial tension [32]. The signs of Poisson's ratio for various combinations in (d) to (f) are indicated. Reprinted with permission from Wiley.

heating, depressurizing, and/or moistening expands the ring such that the rods are drawn inwards.

When arranged as shown in Fig. 1.10(b), the change in size as a consequence of environmental change causes the rods to be pushed out or drawn in such that NTE, NC, and/or NME is/are observed [66].

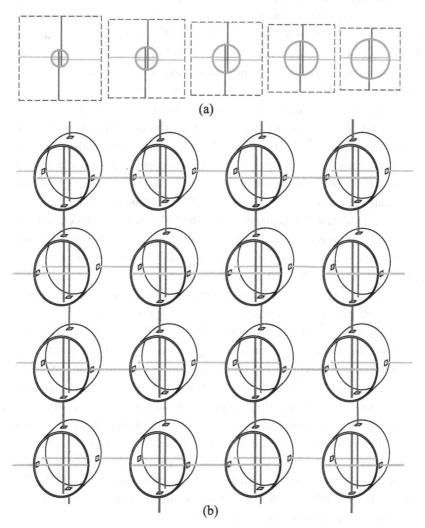

(a)

(b)

Figure 1.10 (a) A unit of ring-rod structure showing unit cell contraction due to draw-in of sliding rods arising from ring expansion (from left to right) or unit cell expansion due to pushing out of rods due to ring contraction (from right to left), and (b) arrangement to prevent turning of rings about any axis [66]. Reprinted with permission from Wiley.

Shortly after the development of the auxetic and NTE/NC/NME metamaterial inspired by the Celtic cross, another mechanical metamaterial was introduced by inspiration from the patriarchal cross (also known as the archiepiscopal cross or *crux gemina*) as well as the Cross of Lorraine. A unit cell of this mechanical metamaterial can be envisaged by a slight modification to the patriarchal/archiepiscopal or to the Lorraine Crosses to produce two slots and two sliders so as to permit sliding connection with its four nearest neighbors that are rotated by 90°, as shown in Fig. 1.11(a). This metamaterial exhibits zero Poisson's ratio when strain is prescribed parallel to the on-axes directions, as depicted in Fig. 1.11(b) and (c). Suppose the unit cells at the original state are arranged in square array and strain is prescribed in the ±45° direction, as indicated in Fig. 1.11(d); then, a Poisson's ratio of $v = -1$ is obtained [277].

Earlier, Cabras and Brun [278] analyzed a series of 2D auxetic systems in which the Poisson's ratio is close to −1 using counterrotating elements. The periodic microstructures investigated are the hexagonal, square, and triangular geometries. For the square geometry, each unit cell consists of a pair of equal-armed crosses pin-joined at their centers. By inspiration from the Iron Cross, this system was modified to manifest NTE and NME characteristics (in addition to exhibiting auxetic properties) if the equal-armed crosses are rigid with insignificant expansion coefficients, while the bridging beams are more compliant with significantly greater expansion coefficients [279].

The incorporation of the bridging beams, as indicated in Fig. 1.12(a), produces a unit cell of the modified metamaterial that partly resembles the symbol of the Iron Cross. The Iron Cross was established by King Frederick William III of Prussia on March 17, 1813 and was originally used as military decoration in the Kingdom of Prussia. From the 2-by-2 square array in its original state illustrated in Fig. 1.12(b), an increase in temperature and/or moisture concentration would lengthen the bridging beams so that the pair of crosses rotates in such a manner that decreases its overall boundary, as indicated in Fig. 1.12(c). By similar reasoning, a decrease in temperature and/or moisture concentration would shorten the bridging beams so that the pair of crosses rotates in such a manner that increases its overall boundary, as indicated in Fig. 1.12(d). The establishment of this so-called Iron Cross metamaterial paved the way for the development of the Maltese Cross metamaterial.

1.4 Maltese Cross Metamaterials

The Maltese Cross is a type of heraldic cross corresponding to the Order of Knights of the Hospital of Saint John (or simply "Knights Hospitaller") and is currently symbolic of the Sovereign Military Order of Malta (SMOM)

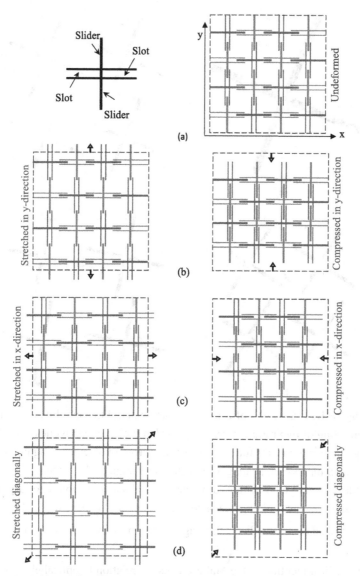

Figure 1.11 (a) A unit cell of the microstructure and 4-by-4 unit cells in the original state, (b) loadings in the *y*-direction, (c) loadings in the *x*-direction, and (d) off-axis loading [277]. Note: the dashed squares indicate the original boundary. Reprinted with permission from MDPI.

and the island nation of Malta. A few salient examples of the Maltese Cross are furnished in Fig. 1.13(a) to (c), while a geometrical construction of the Maltese cross is furnished in Fig.1.13(d) on the basis of a regular octagon.

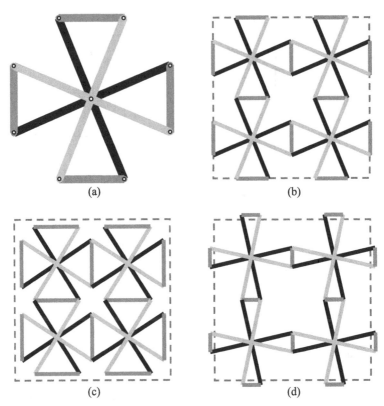

(a)   (b)

(c)   (d)

Figure 1.12 (a) A unit of Iron Cross metamaterial that exhibits $v = -1$ consisting of two rigid crosses pin-joined at their centers, while (b) each bridging beam connects four ends of crosses from two neighboring units. (c) Uniaxial compression, heating, or moistening elongates the bridging beams to produce overall contraction, (d) while uniaxial tension, cooling, or drying shortens the bridging beams to generate overall expansion [279].

Note: Dashed squares indicate the original boundary in (b) for comparison.

The earliest mention of the Maltese Cross in relation to optical/electromagnetic metamaterials is attributed to Zhu et al. [280]. Within the context of mechanical metamaterials, the earliest work inspired by the Maltese Cross was used for designing a 2D material with positive and negative CTE and CME that is tunable [281].

As shown in Fig. 1.14, the Maltese cross consists of two equal-armed crosses pin-joined at their centers to permit relative rotation. The ends of the equal-armed crosses are attached to hinge rods, which are connected to the connecting rods. Suppose the layers of the convex and concave sides of the bimaterial spiral spring possess higher and lower expansion coefficients, respectively; then an

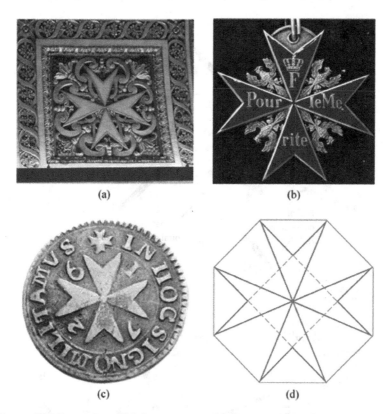

Figure 1.13 Examples of Maltese crosses: (a) Ornamental Maltese cross on the ceiling of St. John's Co-Cathedral, Valletta, Malta (Mattia Preti, 1660s), as photographed by Kritzolina https://commons.wikimedia.org/wiki/File: Detail_in_der_St._John%27s_Co-Cathedral,_Valetta,_Malta_Nov_2014.JPG; (b) *Pour le Mérite* order of merit badge since the eighteenth century https://en.wikipedia.org/wiki/File:Blue_Max.jpg; and (c) a Maltese coin from the early eighteenth century https://www.cngcoins.com/Coin.aspx?CoinID=79406. (d) Geometrical construction of the Maltese cross from a regular octagon.

increase in the environmental temperature and/or moisture concentration would increase the curvature of the spiral spring.

Based on the manner in which the two equal-armed crosses are connected to the spiral spring shown in Fig. 1.14, the increase in the spiral spring's curvature rotates the cross ABCD clockwise to A′B′C′D′, while the cross EFGH rotates counterclockwise to E′F′G′H′ such that the narrowing of the Maltese cross arms pulls the connecting rods radially inward so as to give NTE or NME, or even negative hygrothermal expansion (NHTE) when both NTE and NME are present. The linkage motion reverses when the curvature

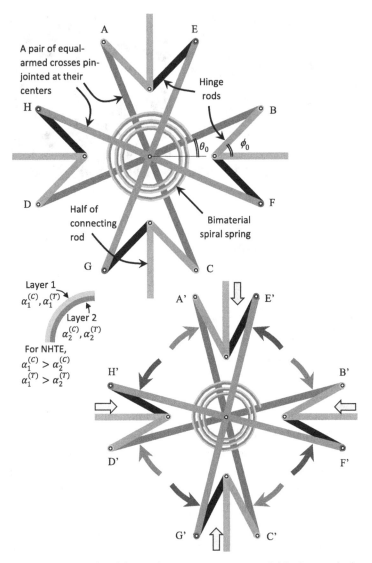

Figure 1.14 A unit of the Maltese cross metamaterial before and after temperature and/or moisture increase. NHTE properties are exhibited based on the manner in which the bimaterial spiral spring is attached to the equal-armed crosses [281]. Reprinted with permission from the Royal Society.

of the bimaterial spiral decreases because of decreasing temperature and/or moisture concentration.

The increased dimension of the metamaterial due to pushout of the connecting rods in conjunction with cooling and/or drying gives rise to NTE and/or NME again. If the convex and the concave layers are exchanged such that the former and latter possess lower and higher expansion coefficients, respectively, then the Maltese cross metamaterial exhibits positive effective expansion coefficients. The observation of strain of equal sign along both axes arising from environmental changes suggests that, under the prescription of uniaxial strain parallel to one of the axes, the resulting strain on the other axis possesses equal sign.

In other words, the Maltese cross metamaterial exhibits auxetic properties regardless of whether it is NTE and NME, or conventional. In fact, the auxeticity is unaffected when the bimaterial spiral spring is replaced by a homogeneous or single-layered spiral spring. The analysis on the auxeticity of this metamaterial is developed and discussed in the subsequent sections.

2 Twofold Symmetrical Mechanism

2.1 The Counterrotating Cross Mechanism

The unit of metamaterial is similar to that shown in Fig. 1.14, that is, it consists of a pair of equal-armed crosses that are pin-jointed at their centers while their vertices are connected in such a manner that the shape of a Maltese cross is manifested, except that the mandated bimaterial strip for the spiral spring is no longer a requirement. It will later be shown that the metamaterial exhibits $v < 0$ in the on-axes directions; the Poisson's ratio in the off-axes directions can be positive or negative, depending on the direction of consideration. In addition to the pair of equal-armed crosses joined at their junctions, each unit of the metamaterial is linked with four neighboring units via connecting rods, as shown in Fig. 2.1. The equal-armed crosses are joined to the connecting rods via hinge rods. For each unit cell, there is a total of 13 pin-joints. The pin-joints are listed here, with reference to Fig. 2.1:

- One pin-joint at the center between the equal-armed cross ABCD and the equal-armed cross EFGH.
- Eight pin-joints between the equal-armed crosses and the hinge rods; these pin-joints are labelled as A, E, B, F, C, G, D, and H.
- Four pin-joints between hinge rods and connecting rods; they are indicated at J, I, N, and M.

When a tensile load is applied on the horizontal connecting rods, as shown by the horizontal arrows in Fig. 2.1, the equal-armed cross ABCD rotates counterclockwise to A'B'C'D', while the equal-armed cross EFGH simultaneously rotates clockwise to E'F'G'H'. As a consequence, the vertical connecting rods move away from the center of the Maltese cross, as shown by the vertical

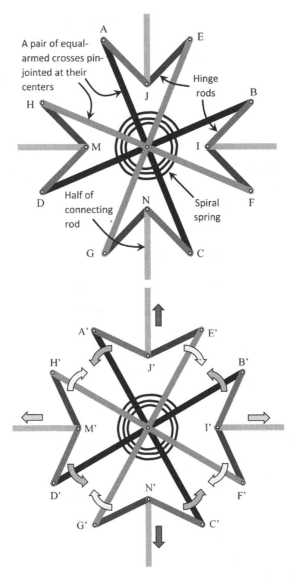

Figure 2.1 A unit of the Maltese cross auxetic metamaterial in its original state, and upon application of uniaxial tension.

arrows. Needless to say, the directions of these motions are reversed when the applied force is compressive. Whether the applied load is tensile or compressive, the on-axes strains are of equal signs, thereby stipulating a negative Poisson's ratio.

To visualize how the translational motions of the connecting rods influence the rotations of the equal-armed crosses and of the hinge rods, we consider four units of the metamaterial arranged in square array, and, at the original state, the equal-armed crosses in each pair are separated by an angle of 45° (see Fig. 2.2 left center) such that the array of metamaterials deform to those shown in Fig. 2.2 (top left) and (bottom left) under prescribed tensile and compressive uniaxial strains, respectively. The metamaterials thereafter deform into those shown in Fig. 2.2 (top right) and (bottom right) upon further prescribed uniaxial strains.

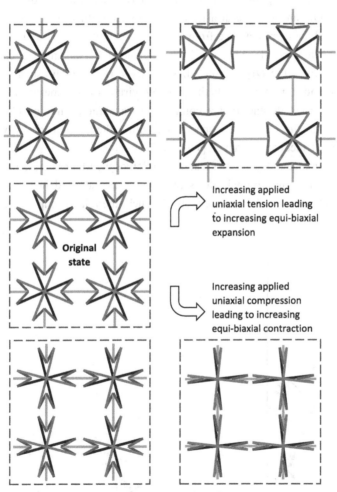

Figure 2.2 A 2-by-2 square array of the metamaterial in its original state (center left) experiencing equi-biaxial strain arising from application of uniaxial tension (top) and idealized uniaxial compression (bottom) based on the twofold symmetrical mechanism. The dashed rectangles indicate the original dimension of the 2-by-2 array. Springs and hinges are not shown for clarity.

2.2 Infinitesimal Deformation

In establishing the overall Poisson's ratio within the context of infinitesimal deformation, the models are governed purely by geometrical descriptions such as l, l_i, and l_j for all the considered modes of deformation. Figure 2.3 shows only a portion of the metamaterial representative unit that is sufficient for the purpose of small deformation analysis, whereby the origin O refers to the pin-joint for the two equal-armed crosses. The portion AOB, being two arms of the ABCD equal-armed cross, is rigid such that OA and OB are at a right angle to each other and subtended by the angle θ counterclockwise from the Ox_2-axis and the Ox_1-axis, respectively. The hinge rods AJ and BI are subtended by an angle ϕ counterclockwise from the Ox_2-axis and the Ox_1-axis, respectively, such that the hinges J and I remain on their axes by virtue of symmetry. Thus the connecting rods IK and JL are confined to translational motion along the Ox_1-axis and the Ox_2-axis, respectively, within the framework of twofold symmetrical mechanism developed in this section.

With reference to the geometrical description of the Maltese cross furnished in Fig. 1.13(d), the original angles of θ and ϕ as indicated in Fig. 2.3 are $\theta_0 = 22.5°$ and $\phi_0 = 45°$, respectively. As such, the relationship between the

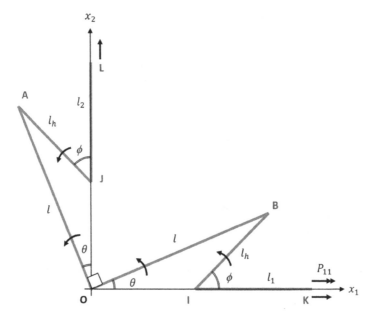

Figure 2.3 Schematic for infinitesimal deformation analysis of the metamaterial according to the twofold symmetrical mechanism.

length for each arm of the equal-armed cross OA or OB, l, can be related to the length of the hinge rod AJ or BI, l_h as

$$l_h \sin\frac{\pi}{4} = l \sin\frac{\pi}{8}. \tag{2.1}$$

Substituting $\sin(\pi/4) = 1/\sqrt{2}$ and $\sin(\pi/8) = \sqrt{2 - \sqrt{2}}/2$ into Eq. (2.1) gives a direct relationship between l_h and l as

$$l_h = \sqrt{\frac{2 - \sqrt{2}}{2}}\, l. \tag{2.2}$$

Whether a uniaxial tensile force of P_{11} or P_{22} is applied along the Ox_1-axis or along the Ox_2-axis, respectively, AOB rotates in the counterclockwise direction such that the angle θ increases. For the connecting rods JL and IK to remain on the axes, the hinge rods AJ and BI must also rotate in the counterclockwise direction such that ϕ increases. Under the action of uniaxial compressive load on either axis, the abovementioned rotational and translational motions reverse such that both θ and ϕ decrease. Although the subsequent analysis considers the application of tensile load, that is, P_{ii} is positive with the changes in angles $d\theta$ and $d\phi$ being positive, the analysis remains valid for the application of compressive force wherein the signs of $d\theta$ and $d\phi$ are reversed.

Projecting the rods OB and BI on the Ox_2-axis, we have $l \sin\theta = l_h \sin\phi$ such that taking its derivative,

$$\frac{d}{d\theta} l \sin\theta = \frac{d}{d\phi} l_h \sin\phi \cdot \left(\frac{d\phi}{d\theta}\right), \tag{2.3}$$

gives $l \cos\theta d\theta = l_h \cos\phi d\phi$. Expressing l_h in terms of l using Eq. (2.2) yields

$$l \cos\theta d\theta = \sqrt{\frac{2 - \sqrt{2}}{2}}\, l \cos\phi d\phi, \tag{2.4}$$

whereupon substituting $\theta = 22.5°$ and $\phi = 45°$ for the original state gives the relationship between the infinitesimal rotational angles as

$$d\phi = \sqrt{\frac{2 + \sqrt{2}}{2 - \sqrt{2}}}\, d\theta. \tag{2.5}$$

Having established the preceding relation, the following develops the overall Poisson's ratio. With reference to Fig. 2.3, the distances OK and OL are set as the original dimensions x_1 and x_2 as measured along the Ox_1-axis and Ox_2-axis, respectively, so that the displacements of points K and L, that is, dx_1 and dx_2, denote the incremental change in the size corresponding to the axes for a quarter unit of the metamaterial. Hence

$$x_i = l \cos\theta_0 - l_h \cos\phi_0 + l_i = \frac{l}{2}\left(\sqrt{2+\sqrt{2}} - \sqrt{2-\sqrt{2}}\right) + l_i \qquad (2.6)$$

and

$$dx_i = l_h d\phi \sin\phi_0 - l d\theta \sin\theta_0 = \frac{\sqrt{2+\sqrt{2}} - \sqrt{2-\sqrt{2}}}{2} l d\theta, \qquad (2.7)$$

where $i,j = 1,2$. The engineering strains can therefore be obtained from Eqs. (2.6) and (2.7) as

$$\varepsilon_{ii} = \frac{dx_i}{x_i} = \frac{\left(\sqrt{2+\sqrt{2}} - \sqrt{2-\sqrt{2}}\right)d\theta}{\sqrt{2+\sqrt{2}} - \sqrt{2-\sqrt{2}} + 2l_i/l}. \qquad (2.8)$$

Therefore the Poisson's ratios for loading in the on-axes directions are

$$v_{ij} = \frac{1}{v_{ji}} = -\frac{\varepsilon_{jj}}{\varepsilon_{ii}} = -\frac{\sqrt{2+\sqrt{2}} - \sqrt{2-\sqrt{2}} + 2l_i/l}{\sqrt{2+\sqrt{2}} - \sqrt{2-\sqrt{2}} + 2l_j/l}. \qquad (2.9)$$

For square array, imposing $l_1 = l_2$ gives

$$v_{ij} = v_{ji} = -1. \qquad (2.10)$$

For rectangular array, $|v_{ij}| > |v_{ji}|$ if $l_i > l_j$.

2.3 Finite Deformation

With reference to the twofold symmetrical deformation mechanism illustrated in Fig. 2.2, a quarter of the unit mechanism linkage is isolated, as shown in Fig. 2.4 for analysis, whereby the original linkages OA and OB rotate counterclockwise by $\Delta\theta$ to OA' and OB', respectively, under the influence of uniaxial tensile force P_{11} or P_{22} such that the subtended angle with respect to their corresponding axes changes from θ_0 to θ. Simultaneously, AJ and BI rotate counterclockwise by an angle $\Delta\phi$ from ϕ_0 to ϕ with reference to Ox_2- and Ox_1-axes, respectively, while JL and IK undergo translational motion to J'L' and I'K', respectively, as indicated in Fig. 2.4. The positions of K' and L', therefore, indicate the updated dimensions x_1' and x_2', respectively, for a quarter unit of the metamaterial.

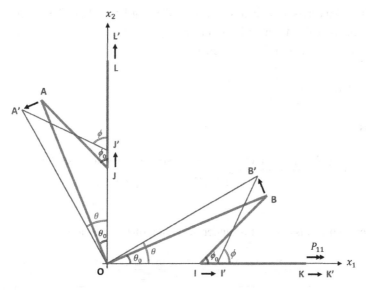

Figure 2.4 Schematic for finite deformation analysis of the metamaterial within the framework of the twofold symmetrical mechanism.

The finite rotation relationship between $\Delta\theta$ and $\Delta\phi$ is sought by projecting OB' and $B'I'$ against the Ox_2-axis so as to equate their Ox_2-component lengths,

$$l \sin(\theta_0 + \Delta\theta) = \sqrt{\frac{2 - \sqrt{2}}{2}} l \sin(\phi_0 + \Delta\phi), \tag{2.11}$$

to yield

$$\cos \Delta\theta + \sqrt{\frac{2 + \sqrt{2}}{2 - \sqrt{2}}} \sin \Delta\theta = \cos \Delta\phi + \sin \Delta\phi. \tag{2.12}$$

With reference to Fig. 2.4, wherein

$$\begin{aligned} \theta &= \theta_0 + \Delta\theta \\ \phi &= \phi_0 + \phi\Delta \end{aligned} \tag{2.13}$$

the positions of K' $(i = 1)$ and L' $(i = 2)$, that is,

$$x_i = l \cos \theta - l_h \cos \phi + l_i, \tag{2.14}$$

can be written as

$$x_i' = \frac{l}{2}\left[\sqrt{2 + \sqrt{2}} \cos \Delta\theta - \sqrt{2 - \sqrt{2}}(\sin \Delta\theta + \cos \Delta\phi - \sin \Delta\phi)\right] + l_i, \tag{2.15}$$

in which $\Delta\phi$ can be obtained for a given $\Delta\theta$ using Eq. (2.12). The incremental strain $d\varepsilon$ is defined similarly as the engineering strain so that, taking into account the finite deformation, we have the logarithmic strain

$$\varepsilon = \int d\varepsilon = \ln\left(\frac{x_i'}{x_i}\right), \tag{2.16}$$

that is,

$$\varepsilon_{ii} = \ln \frac{\sqrt{2+\sqrt{2}}\cos\Delta\theta - \sqrt{2-\sqrt{2}}(\sin\Delta\theta + \cos\Delta\phi - \sin\Delta\phi) + 2l_i/l}{\sqrt{2+\sqrt{2}} - \sqrt{2-\sqrt{2}} + 2l_i/l}. \tag{2.17}$$

Therefore the Poisson's ratios under the action of P_{ii} or P_{jj} are

$$v_{ij} = \frac{1}{v_{ji}} = -\frac{\ln \dfrac{\sqrt{2+\sqrt{2}}\cos\Delta\theta - \sqrt{2-\sqrt{2}}(\sin\Delta\theta + \cos\Delta\phi - \sin\Delta\phi) + 2l_j/l}{\sqrt{2+\sqrt{2}} - \sqrt{2-\sqrt{2}} + 2l_j/l}}{\ln \dfrac{\sqrt{2+\sqrt{2}}\cos\Delta\theta - \sqrt{2-\sqrt{2}}(\sin\Delta\theta + \cos\Delta\phi - \sin\Delta\phi) + 2l_i/l}{\sqrt{2+\sqrt{2}} - \sqrt{2-\sqrt{2}} + 2l_i/l}} \tag{2.18}$$

Again, Eq. (2.10) is recovered upon substitution of $l_i = l_j$ for square array.

2.4 Results and Limitations

Figure 2.5 shows the rotational angle relationship between the equal-armed crosses and the hinged rods based on the assumption of very small strain described by Eq. (2.5) and on the more accurate model set out in Eq. (2.12) for uniaxial stretching and for the symmetrically idealized uniaxial compression. For a given infinitesimal rotation $d\theta$, the approximate model underestimates the infinitesimal rotational angle $d\phi$ under uniaxial tensile loading but overestimates the magnitude under uniaxial compressive loading. The good agreement observed for small rotation of equal-armed cross is not surprising since Eq. (2.5) is recovered from Eq. (2.12) by applying the small angular change approximations.

The effect of the metamaterial's spacing on the Poisson's ratio v_{ij} based on uniaxial stretching (Fig. 2.2, top) and symmetrically idealized uniaxial compression (Fig. 2.2, bottom) for prescribed strain ε_{ii}, in terms of dimensionless lengths of the connecting rods, is shown in Fig. 2.6(a) for various $2l_i/l$ at fixed $2l_j/l$ and in Fig. 2.6(b) for various $2l_j/l$ at fixed $2l_i/l$. Note that, since l_1 and l_2 represent the half-lengths of the connecting rods aligned parallel to the Ox_1 and Ox_2 axes, respectively, the terms $2l_i/l$ and $2l_j/l$ represent the dimensionless forms of the connecting rods' full lengths. Plotted results using Eq. (2.18) show that the exact model for the Poisson's ratio is only marginally influenced by the

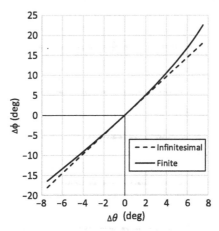

Figure 2.5 Relationship between the hinge rod rotation and the equal-armed cross rotation.

applied strain, while the Poisson's ratio according to the approximate model described by Eq. (2.9) is not only unaffected by the applied strain, but also provides very good agreement. This good agreement between the approximate and exact models can be shown by implementing small angular change approximations on Eq. (2.17) and using the Maclaurin series approximation for the natural logarithm to suggest the following expression upon truncating higher-order terms:

$$\varepsilon_{ii} \approx \ln\left(1 + \frac{\sqrt{2-\sqrt{2}}(d\phi - d\theta)}{\sqrt{2+\sqrt{2}} - \sqrt{2-\sqrt{2}} + 2l_i/l}\right)$$

$$\approx \frac{\sqrt{2-\sqrt{2}}(d\phi - d\theta)}{\sqrt{2+\sqrt{2}} - \sqrt{2-\sqrt{2}} + 2l_i/l}. \tag{2.19}$$

Equation (2.8) is recovered when $d\phi$ on the right-hand side of Eq. (2.19) is expressed in terms of $d\theta$ using Eq. (2.5), thereby confirming the validity of the Poisson's ratio described in Eq. (2.9) for small deformation.

It needs to be cautioned that, while the deformation mechanism for on-axis uniaxial stretching illustrated in Fig. 2.2 (top) is physically admissible, that for on-axis uniaxial compression shown in Fig. 2.2 (bottom) is highly idealized. Since the equal-armed crosses in each pair are held by a spiral spring while the connecting rods are permitted to rotate with reference to the hinge rods, each Maltese cross behaves as a rigid body under non-tensile loads. It is therefore more likely in reality that, due to any slight asymmetry or imperfection, the

(a)

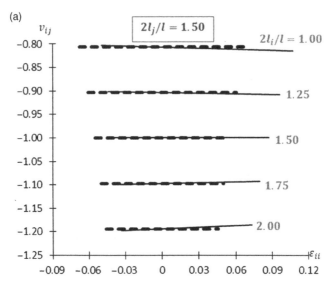

(b)

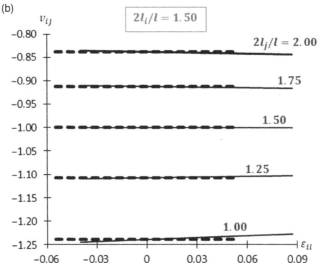

Figure 2.6 Effect of metamaterial spacing on Poisson's ratio v_{ij} under prescribed strain ε_{ii} for (a) various l_i at fixed l_j and (b) various l_j at fixed l_i, based on uniaxial stretching and symmetrically idealized uniaxial compression illustrated in Fig. 2.2. The dashed lines and continuous curves denote the infinitesimal and finite models, respectively.

prescription of on-axis uniaxial compression brings about the rotation of the Maltese cross as a rigid body.

Consider again the 2-by-2 square array of this metamaterial from its original state in Fig. 2.7(a), in which the Maltese crosses rotate in one direction akin to the tetrachiral deformation while the connecting rods rotate in the opposite direction as shown in Fig. 2.7(b), thereby retaining its auxetic behavior during uniaxial compression, albeit undergoing a different mechanism route. Alternatively, all the horizontal and vertical connecting rods remain parallel to the two axes, while the Maltese crosses rotate as shown in Fig. 2.7(c). This mechanism resembles the anti-tetrachiral deformation. Suppose the load is applied along $\pm 45°$ to the axes or an applied shear strain of $\varepsilon_{12} = \varepsilon_{21}$ is prescribed; then overall shearing occurs as shown in Fig. 2.7(d). In all these three cases of mechanism, the Maltese crosses rotate as rigid bodies, while the connecting rods undergo rotation relative to the Maltese crosses. In other words, the metamaterial exhibits partial auxeticity because it behaves as an auxetic system as shown in Fig. 2.2 (top and bottom) as well as Fig. 2.7(b) and (c), but as a non-auxetic system in Fig. 2.7(d). In general, "partial auxeticity" refers to anisotropic solids whereby the Poisson's ratio can range from positive to negative, depending on the direction of the applied uniaxial stress. In the case of 2D materials whereby loading is applied in the material plane, partial auxeticity can be mathematically defined as

$$
\begin{aligned}
v < 0 &\iff \varsigma_0 < \varsigma < \varsigma_0 + \pi \\
v > 0 &\iff \varsigma_0 - \pi < \varsigma < \varsigma_0
\end{aligned}
\tag{2.20}
$$

where ζ is the angle measured counterclockwise from the Ox_1 axis, with ζ_0 being the loading direction that gives $v = 0$.

A partially auxetic material with auxetic property in the on-axes direction but with non-auxetic behavior, and hence low shear modulus, in the $\pm 45°$ directions has practical applications – for example, those in which high indentation resistance and unique shape change that are possible with negative Poisson's ratio but low shear modulus relative to Young's modulus are desirable (such as in helmets or blast paneling). Some of these applications have been discussed by Duncan and colleagues [236], Evans and Alderson [282], Cheng and colleagues [283], and Bliven and colleagues [284].

Out of the three possible modes of deformation mechanism under on-axis uniaxial compression, the chiral-like mechanisms depicted in Fig. 2.7(b) and (c) have greater propensity than the symmetrically idealized twofold symmetry (or fourfold symmetry for square array). This is because the deformation modes of lower stiffness are dominant over the deformation modes of higher stiffness. This effect has been observed experimentally on missing-rib [285] and tetra-chiral honeycomb [286] models, which have the potential to exhibit auxetic behavior and have a very low shear modulus. If during loading these systems are constrained to retain their alignment with the x- and y-axes, then they exhibit auxetic behavior [287]. Suppose the metamaterial deforms in the manner shown

Figure 2.7 Schematics of a 2-by-2 metamaterial unit (a) in its original state, (b) a tetrachiral mode of deformation via uniform rotation of Maltese crosses as a possible mechanism during applied compressive loading, (c) an anti-tetrachiral mode of deformation via counterrotation of the Maltese crosses as an alternative mechanism during applied compressive loading, and (d) a shearing mode of deformation with reference to the axes during off-axis loading.

in Fig. 2.2 (top) during uniaxial stretching and either Fig. 2.7(b) or Fig. 2.7(c) instead of Fig. 2.2 (bottom) during prescription of uniaxial compression; then only the positive parts of Fig. 2.5 and the tensile strain parts of Fig. 2.6 are valid, while the negative part is undefined since the Maltese crosses are more likely to rotate as rigid bodies such that θ and ϕ remain constant during compression.

While the twofold symmetrical deformation mechanism is likely to occur during applied on-axis tension and the shear mechanism is likely to take place during pure shearing and off-axis tensile or compressive loading, there are various deformation mechanisms that may possibly take place as a result of applied on-axis compression. For the sake of systematic analysis, all the possible modes of deformation mechanisms are separately established in the subsequent sections. This does not, however, imply that all the mechanisms to be discussed are equally likely to occur upon loading. The various deformation mechanisms will be revisited in the final section in order to appraise the comparative likelihood that each mechanism will manifest, as well as the conditions that are required for the least likely mechanism to take place.

3 Tetrachiral Mechanism

3.1 Chiral Mechanism

One group of microstructural geometries that gives rise to auxetic behavior is the chiral honeycombs. A few examples of chiral honeycombs are furnished in Fig. 3.1, namely the hexachiral, the tetrachiral, and the trichiral honeycombs. The chiral honeycombs are named as such due to the uniform rotation of the nodes (indicated as circles in Fig. 3.1) in response to in-plane stretching; the rotation of the nodes is uniformly reversed when the in-plane force is compressive. The prefixes denote the number of connections each node makes with its nearest neighbors, hence "hexa-," "tetra-," and "tri-" denote six, four, and three connections, respectively. The reader is referred to the review by Wu and colleagues [238] for further details on this class of auxetic honeycombs.

The manner in which the chiral honeycombs deform may well be illustrated by the example of tetrachiral deformation described in Fig. 3.2. Consider a 3-by-3 square array of tetrachiral honeycomb in its original state shown in Fig. 3.2 (top left), whereby the ribs are slightly curved and protrude from the nodes in the counterclockwise direction. Suppose a uniaxial load is applied parallel to the Ox_1-direction; the nodes not only displace further from each other in the Ox_1-direction, but also uniformly rotate in the counterclockwise direction while straightening the ribs. The straightening of the ribs also causes the nodes to be displaced further from each other along the Ox_2-direction, as shown in Fig. 3.2 (top right), thereby manifesting auxetic behavior. When the in-plane load is reversed, the nodes are displaced closer to each other while the node

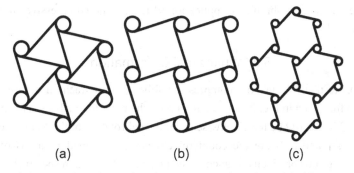

(a)   (b)   (c)

Figure 3.1 Examples of auxetic microstructures based on chiral honeycombs: hexachiral (a), tetrachiral (b), and trichiral (c).

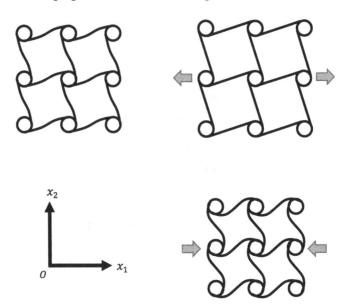

Figure 3.2 Demonstration of auxetic behavior in tetrachiral honeycomb under uniaxial loads.

rotation is uniformly reversed, which results in increased curvature of the ribs. The curving of the ribs decreases the node spacing in the Ox_2-direction, as indicated in Fig. 3.2 (bottom right), thereby exhibiting auxetic characteristics again. The deformation pattern of the tetrachiral honeycomb under compressive load shares some similarities to that shown in Fig. 2.7(b). For this reason, the mechanism shown in Fig. 2.7(b) has been termed the "tetrachiral" mechanism. It can be seen from Fig. 3.2 that, when undergoing tetrachiral mechanism, the metamaterial has potential application as a sieve with stress-controlled pore size, that is, the spaces confined by the linkages act as pores wherein their size can be enlarged or shrunk by application of tensile and compressive stresses, respectively.

3.2 Infinitesimal Deformation

Supposing that the uniaxial compression follows the tetrachiral deformation pattern furnished in Fig. 2.7(b), the required schematic for analysis is as shown in Fig. 3.3. Let $d\Omega$ be the clockwise infinitesimal rotation of the Maltese cross while $d\Omega_1$ and $d\Omega_2$ are the counterclockwise infinitesimal rotations of the connecting rods originally aligned to the Ox_1-and Ox_2-axes, respectively; then the updated dimensions for x_1 and x_2 are

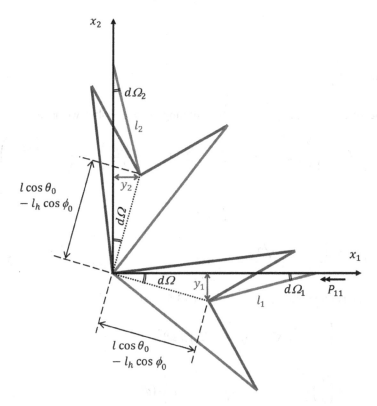

Figure 3.3 A portion of the metamaterial unit for analysis of the tetrachiral mechanism under uniaxial compression illustrated in Fig. 2.7(b).

$$x_i' = x_i + dx_i = (l\cos\theta - l_h\cos\phi)\cos d\Omega + l_i\cos d\Omega_i \qquad (3.1)$$

for $i = 1, 2$.

The rotational angles $d\Omega_i$ can be related to $d\Omega$ with reference to y_i ($i = 1, 2$) to give

$$\sin d\Omega_i = \left(\frac{l}{l_i}\cos\theta - \frac{l_h}{l_i}\cos\phi\right)\sin d\Omega, \qquad (3.2)$$

in which $y_1 = y_2$. Implementing Taylor series for small angular rotation, Eqs. (3.1) and (3.2) reduce to

$$x_i' = x_i + dx_i = (l\cos\theta - l_h\cos\phi)\left(1 - \frac{(d\Omega)^2}{2}\right) + l_i\left(1 - \frac{(d\Omega_i)^2}{2}\right)$$

$$(3.3)$$

and

$$d\Omega_i = \left(\frac{l}{l_i} \cos \theta - \frac{l_h}{l_i} \cos \phi \right) d\Omega \qquad (3.4)$$

respectively. Using Eq. (2.2) and substituting $2\theta = \phi = 45°$, we have

$$x_i + dx_i = \frac{l}{2} \left(\sqrt{2 + \sqrt{2}} - \sqrt{2 - \sqrt{2}} \right) \left(1 - \frac{(d\Omega)^2}{2} \right) + l_i \left(1 - \frac{(d\Omega_i)^2}{2} \right) \qquad (3.5)$$

and

$$d\Omega_i = \frac{\sqrt{2 + \sqrt{2}} - \sqrt{2 - \sqrt{2}}}{2} \frac{l}{l_i} d\Omega \qquad (3.6)$$

such that subtracting Eq. (2.6) from Eq. (3.5) gives

$$dx_i = -\frac{l}{4} \left(\sqrt{2 + \sqrt{2}} - \sqrt{2 - \sqrt{2}} \right) (d\Omega)^2$$

$$- \left(\frac{1}{8} \right) \left(\sqrt{2 + \sqrt{2}} - \sqrt{2 - \sqrt{2}} \right)^2 \frac{l^2}{l_i} (d\Omega)^2 \qquad (3.7)$$

and hence the infinitesimal strain

$$\varepsilon_{ii} = \frac{dx_i}{x_i} = \frac{\left(\sqrt{2 + \sqrt{2}} - \sqrt{2 - \sqrt{2}} \right) \left[1 + \frac{\sqrt{2+\sqrt{2}} - \sqrt{2-\sqrt{2}}}{2} \frac{l}{l_i} \right] \left[-\frac{(d\Omega)^2}{2} \right]}{\sqrt{2 + \sqrt{2}} - \sqrt{2 - \sqrt{2}} + 2l_i/l}. \qquad (3.8)$$

Therefore, the overall infinitesimal Poisson's ratio under uniaxial compression is

$$v_{ij} = \frac{1}{v_{ji}} = -\frac{\varepsilon_{jj}}{\varepsilon_{ii}} = -\frac{\sqrt{2 + \sqrt{2}} - \sqrt{2 - \sqrt{2}} + \frac{2l_i}{l}}{\sqrt{2 + \sqrt{2}} - \sqrt{2 - \sqrt{2}} + \frac{2l_i}{l}} \frac{\left[1 + \frac{\sqrt{2+\sqrt{2}} - \sqrt{2-\sqrt{2}}}{2} \frac{l}{l_j} \right]}{\left[1 + \frac{\sqrt{2+\sqrt{2}} - \sqrt{2-\sqrt{2}}}{2} \frac{l}{l_i} \right]}$$

$$= -\frac{l_i}{l_j}. \qquad (3.9)$$

Although a convenient expression for the Poisson's ratio is obtained, the alternative expression is useful for comparing against Eq.(2.9) for the case of twofold symmetrical deformation. As with the infinitesimal twofold symmetrical deformation, the Poisson's ratio reduces to Eq. (2.10) for square array, while $|v_{ij}| > |v_{ji}|$ for rectangular array with $l_i > l_j$.

3.3 Finite Deformation and Maximum Rotation

Supposing again the uniaxial compression obeys the tetrachiral deformation mechanism indicated in Fig. 2.7(b) instead of the symmetrically idealized deformation denoted in Fig. 2.2 (bottom), then in the case of finite deformation, Eqs. (3.1) and (3.2) are not simplified to Eqs. (3.3) and (3.4) respectively. Instead, the substitution of Eq. (3.2) into Eq. (3.1) gives

$$x_i' = (l\cos\theta - l_h\cos\phi)\cos\Delta\Omega$$

$$+ l_i \cos\left\{\sin^{-1}\left[\left(\frac{l}{l_i}\cos\theta - \frac{l_h}{l_i}\cos\phi\right)\sin\Delta\Omega\right]\right\} \tag{3.10}$$

such that, considering $2\theta = \phi = 45°$ for rigid Maltese cross and using Eq. (2.2), the true strain can be expressed as

$$\varepsilon_{ii} =$$

$$\ln\frac{\left(\sqrt{2+\sqrt{2}} - \sqrt{2-\sqrt{2}}\right)\cos\Delta\Omega + 2l_i/l\cos\left\{\sin^{-1}\left[\frac{\sqrt{2+\sqrt{2}}-\sqrt{2-\sqrt{2}}}{2}\frac{l}{l_i}\sin\Delta\Omega\right]\right\}}{\sqrt{2+\sqrt{2}} - \sqrt{2-\sqrt{2}} + 2l_i/l}. \tag{3.11}$$

This gives the overall finite Poisson's ratio:

$$v_{ij} =$$

$$-\ln\frac{\left(\sqrt{2+\sqrt{2}} - \sqrt{2-\sqrt{2}}\right)\cos\Delta\Omega + 2l_j/l\cos\left\{\sin^{-1}\left[\frac{\sqrt{2+\sqrt{2}}-\sqrt{2-\sqrt{2}}}{2}\frac{l}{l_j}\sin\Delta\Omega\right]\right\}}{\sqrt{2+\sqrt{2}} - \sqrt{2-\sqrt{2}} + 2l_j/l}$$

$$\overline{\ln\frac{\left(\sqrt{2+\sqrt{2}} - \sqrt{2-\sqrt{2}}\right)\cos\Delta\Omega + 2l_i/l\cos\left\{\sin^{-1}\left[\frac{\sqrt{2+\sqrt{2}}-\sqrt{2-\sqrt{2}}}{2}\frac{l}{l_i}\sin\Delta\Omega\right]\right\}}{\sqrt{2+\sqrt{2}} - \sqrt{2-\sqrt{2}} + 2l_i/l}} \tag{3.12}$$

The maximum rotation, $\Delta\Omega_{max}$, is defined when the connecting rods are aligned to the hinge rods. This limit is determined by the shorter of the two sets of connecting rods. An example is demonstrated in Fig. 3.4, in which $l_1 < l_2$, and so the rotation $\Delta\Omega_{max}$ is limited by l_1. Let l_1 be the lower of the two half-lengths of the connecting rods in the subsequent analysis for describing $d\Omega_{max}$. Note

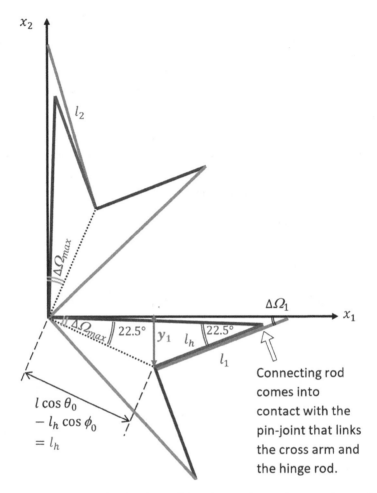

Figure 3.4 Schematics for determining the maximum free-rotation angle during tetrachiral compression.

that the hinge rod is originally aligned at 45° to the Ox_1 axis, and the entire Maltese cross rotates as a rigid body by an angle $\Delta\Omega_{max}$, during which the hinge rod's angle to the Ox_1 axis reduces until it forms an angle of $\Delta\Omega_1$ to the Ox_1 axis. Hence

$$\Delta\Omega_1 = 45° - \Delta\Omega_{max}. \tag{3.13}$$

With reference to y_1, we have

$$l_1 \sin\Delta\Omega_1 = (l\cos\theta_0 - l_h\cos\phi_0)\sin\Delta\Omega_{max} \tag{3.14}$$

Since $l\cos\theta_0 - l_h\cos\phi_0 = l_h$ when $2\theta_0 = \phi_0 = 45°$, we have

$$l_1 \sin \Delta\Omega_1 = l_h \sin \Delta\Omega_{max}. \tag{3.15}$$

Eliminating $\Delta\Omega_1$ from Eqs. (3.13) and (3.15) leads to the maximum rotation for a given ratio of l/l_1:

$$\cot \Delta\Omega_{max} = 1 + \sqrt{2}\frac{l_h}{l_1} = 1 + \sqrt{2 - \sqrt{2}}\frac{l}{l_1}. \tag{3.16}$$

By similar reasoning, the maximum rotation for the tetrachiral mechanism during on-axis compression is

$$\cot \Delta\Omega_{max} = 1 + \sqrt{2 - \sqrt{2}}\frac{l}{l_2} \tag{3.17}$$

if $l_2 < l_1$.

3.4 Results

Since it has been acknowledged that the compressive mode of deformation in Fig. 2.2 (bottom) applies only for idealized twofold symmetrical deformation, the analytical model for the more realistic tetrachiral furnished in Fig. 2.7(b) has been developed. To visualize how the overall Poisson's ratio varies with geometry and the prescribed compressive strain, Fig. 3.5 shows the plots of infinitesimal v_{ij} (thick dashed lines) and finite v_{ij} (thin continuous curves) versus ε_{ii} for various $1 \leq 2l_i/l \leq 2$ at fixed $2l_j/l = 1.5$ (a) and various $1 \leq 2l_j/l \leq 2$ at fixed $2l_i/l = 1.5$ (b) as a result of tetrachiral compression. The limiting condition for rigid-body rotation of the Maltese cross described by Eq. (3.16) and Eq. (3.17) for tetrachiral compression with $l_1 \leq l_2$ and $l_1 \geq l_2$, respectively, appears in Fig. 3.5 as a boundary that connects the exact Poisson's ratio at the maximum compressive strain. These results reveal that the geometry (e.g. l, l_h, l_1, l_2) plays a greater role than the strain and the associated rotation $d\Omega$.

4 Anti-tetrachiral Mechanism

4.1 Anti-chiral Mechanisms

The anti-chiral mechanism is an alternative deformation mode to the chiral mechanism. As with the chiral honeycombs, anti-chiral honeycombs have also been shown to possess auxetic behavior. Examples of anti-tetrachiral honeycombs are displayed in Fig. 4.1, whereby the anti-tetrachiral and the anti-trichiral honeycombs are analogous to, and therefore counterparts of, the tetrachiral and trichiral honeycombs, respectively. In the case of chiral

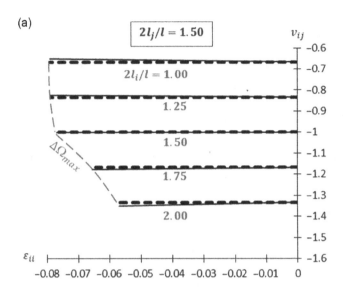

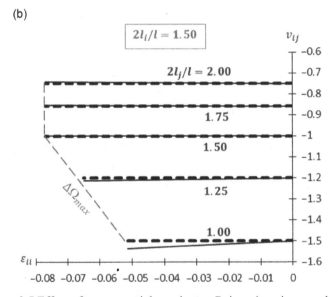

Figure 3.5 Effect of metamaterial spacing on Poisson's ratio v_{ij} under prescribed strain $\varepsilon_{ii} < 0$ for (a) various l_i at fixed l_j and (b) various l_j at fixed l_i based on a tetrachiral mechanism. The dashed lines and continuous curves denote the infinitesimal and finite models, respectively, while the thin dashed lines indicate the maximum rotation defined in Fig. 3.4.

honeycombs, all the ribs protruded from the nodes are in the same direction, namely either all clockwise or all counterclockwise. For the case of anti-chiral honeycombs, the direction of rib protrusion alternates between clockwise and

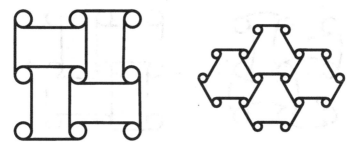

Figure 4.1. Examples of auxetic microstructures based on anti-chiral honeycombs: anti-tetrachiral (a) and anti-trichiral (b).

counterclockwise from one node to the next. As a result, there is no equivalence of the anti-chiral structure for the hexachiral honeycomb; that is, due to geometrical reasons, the anti-hexachiral honeycomb does not exist based on the nomenclature adopted herein. For more details on the comparison between chiral and anti-chiral honeycombs with special emphasis on their auxetic properties, the reader is referred to the works of Alderson et al. [287].

A visual description of how the anti-chiral honeycombs produce auxetic behavior is furnished in Fig. 4.2 using the example of anti-tetrachiral honeycomb. With reference to a 3-by-3 anti-tetrachiral honeycomb in its original state, shown in Fig. 4.2 (top left) with a slight curvature in the ribs, the application of in-plane uniaxial tension increases the spacing of the nodes along the Ox_1-direction, which is facilitated by rotation of the nodes and a simultaneous straightening of the ribs. The straightening of the ribs aligned to the Ox_2-direction causes the nodes to be spaced out in that direction, as indicated in Fig. 4.2 (top right), thereby bringing about auxetic behavior. Unlike the tetrachiral honeycombs, the nodes for anti-tetrachiral honeycombs rotate in both directions such that the rotational direction of a node is opposite to the rotational direction of its four nearest neighboring nodes. When the in-plane load is reversed, the rotational directions of every node are also reversed in addition to the nodes being brought closer together in the Ox_1-direction, such that the ribs are bent to a greater extent. The resulting increase in the curvature of the ribs that are aligned parallel to the Ox_2-direction decreases the spacing between the nodes in that direction, as denoted in Fig. 4.2 (bottom right), thereby exhibiting auxetic characteristics again. Unlike the chiral honeycombs whereby the ribs are bent into S-shaped curves, the ribs in the case of anti-chiral honeycombs are transformed into C-shaped ones. The deformation mechanism of the anti-tetrachiral honeycomb under compressive load shares some common characteristics with those displayed in Fig. 2.7(c). This explains why the mechanism described in

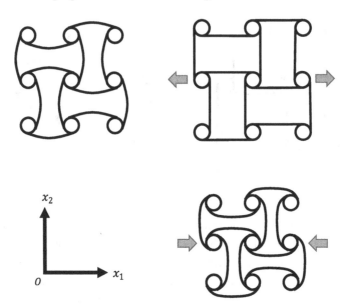

Figure 4.2 Demonstration of auxetic behavior in anti-tetrachiral honeycomb under uniaxial loads.

Fig. 2.7(c) has been called the "anti-tetrachiral" mechanism. Perusal of Fig. 4.2 also indicates that the metamaterial can function as a sieve that can be used for filtering larger particles and particles of various shapes when a uniaxial tension is applied. However, when a uniaxial compression is applied, the sieve not only filters out smaller particles but also permits smaller rod-like particles to go through; the latter is understood to be rod-like particles being filtered (when the metamaterial is under compression) whose lengths are smaller than those of the filtered spherical and regular polyhedral particles when the metamaterial is stretched.

4.2 Infinitesimal Deformation

Supposing the uniaxial compression follows the anti-tetrachiral deformation pattern furnished in Fig. 2.7(c), then the connecting rods of half-lengths l_1 and l_2 displace from, but remain parallel to, the Ox_1 and Ox_2 axes, respectively, as indicated in Fig. 4.3, with $y_1 = y_2$. Substituting $d\Omega_1 = d\Omega_2 = 0$ into Eq. (3.1) gives

$$x_i' = x_i + dx_i = (l\cos\theta - l_h\cos\phi)\cos d\Omega + l_i. \tag{4.1}$$

Imposing the Taylor series approximation

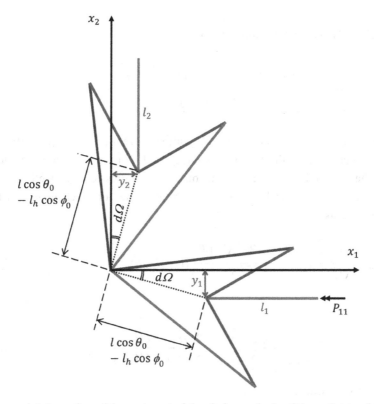

Figure 4.3 A portion of the metamaterial unit for analysis of the anti-tetrachiral mechanism under uniaxial compression illustrated in Fig. 2.7(c).

$$\cos d\Omega = 1 - \frac{1}{2}(d\Omega)^2 + \ldots \tag{4.2}$$

for small angular rotation, Eq. (4.1) simplifies to

$$x'_i = x_i + dx_i = (l \cos \theta - l_h \cos \phi)\left(1 - \frac{(d\Omega)^2}{2}\right) + l_i. \tag{4.3}$$

Substituting Eq. (2.2) and $2\theta = \phi = 45°$ again gives

$$x_i + dx_i = \frac{l}{2}\left(\sqrt{2 + \sqrt{2}} - \sqrt{2 - \sqrt{2}}\right)\left(1 - \frac{(d\Omega)^2}{2}\right) + l_i. \tag{4.4}$$

Subtracting Eq. (2.6) from Eq. (4.4) gives

$$dx_i = -\frac{l}{4}\left(\sqrt{2 + \sqrt{2}} - \sqrt{2 - \sqrt{2}}\right)(d\Omega)^2, \tag{4.5}$$

which leads to the infinitesimal strain

$$\varepsilon_{ii} = \frac{dx_i}{x_i} = \frac{\left(\sqrt{2+\sqrt{2}} - \sqrt{2-\sqrt{2}}\right)\left[-\dfrac{(d\Omega)^2}{2}\right]}{\sqrt{2+\sqrt{2}} - \sqrt{2-\sqrt{2}} + 2l_i/l},$$ (4.6)

hence giving the same Poisson's ratio expression as Eq. (2.9). In other words, the infinitesimal strain ε_{ii} and Poisson's ratio v_{ij} models for the twofold symmetrical tension and compression, as well as the compressive tetrachiral and anti-tetrachiral modes of mechanism can be combined as

$$\varepsilon_{ii} = \frac{\left(\sqrt{2+\sqrt{2}} - \sqrt{2-\sqrt{2}}\right)\Psi}{\sqrt{2+\sqrt{2}} - \sqrt{2-\sqrt{2}} + 2l_i/l}[1 + f(l/l_i)]$$ (4.7)

and

$$v_{ij} = -\frac{\sqrt{2+\sqrt{2}} - \sqrt{2-\sqrt{2}} + 2l_i/l\,[1 + f(l/l_j)]}{\sqrt{2+\sqrt{2}} - \sqrt{2-\sqrt{2}} + 2l_j/l\,[1 + f(l/l_i)]}$$ (4.8)

respectively, where

$$f(l/l_i) = \begin{cases} 0 & \Leftarrow \text{ 2-fold sym.} \\ 0.5\left(\sqrt{2+\sqrt{2}} - \sqrt{2-\sqrt{2}}\right)l/l_i & \Leftarrow \text{ tetrachiral} \\ 0 & \Leftarrow \text{ anti-tetrach.} \end{cases}$$ (4.9)

and the generic angular change symbol ψ refers to

$$\Psi = \begin{cases} d\theta & \Leftarrow \text{ 2-fold sym.} \\ -(d\Omega)^2/2 & \Leftarrow \text{ tetrachiral} \\ -(d\Omega)^2/2 & \Leftarrow \text{ anti-tetrach.} \end{cases},$$ (4.10)

in which the change in the angle of the Maltese cross arm $d\theta$ is positive and negative during stretching and compressing, respectively, while $d\Omega$ refers to the magnitude of the Maltese cross rotation during the tetrachiral and anto-tetrachiral mechanisms, which are the more likely modes of mechanism during compression. If the Maltese crosses are arranged in square array, that is, the half-lengths of the connecting rods aligned along both axes are equal ($l_1 = l_2$), then the Poisson's ratio greatly simplifies to Eq. (2.10) regardless of the mechanism mode.

4.3 Finite Deformation and Maximum Rotation

If the deformation mode under the application of uniaxial compression leads to the anti-tetrachiral mechanism illustrated in Fig. 2.7(c), then substituting Eq. (2.2), $\Delta\Omega_i = 0$, and $2\theta = \phi = 45°$ into Eq. (3.1) gives

$$x_i' = \frac{l}{2}\left(\sqrt{2+\sqrt{2}} - \sqrt{2-\sqrt{2}}\right)\cos\Delta\Omega + l_i, \tag{4.11}$$

which leads to the true strain

$$\varepsilon_{ii} = \ln\frac{\left(\sqrt{2+\sqrt{2}} - \sqrt{2-\sqrt{2}}\right)\cos\Delta\Omega + 2l_i/l}{\sqrt{2+\sqrt{2}} - \sqrt{2-\sqrt{2}} + 2l_i/l} \tag{4.12}$$

and hence the Poisson's ratio

$$v_{ij} = -\frac{\ln\dfrac{\left(\sqrt{2+\sqrt{2}}-\sqrt{2-\sqrt{2}}\right)\cos\Delta\Omega+2l_j/l}{\sqrt{2+\sqrt{2}}-\sqrt{2-\sqrt{2}}+2l_j/l}}{\ln\dfrac{\left(\sqrt{2+\sqrt{2}}-\sqrt{2-\sqrt{2}}\right)\cos\Delta\Omega+2l_i/l}{\sqrt{2+\sqrt{2}}-\sqrt{2-\sqrt{2}}+2l_i/l}}. \tag{4.13}$$

Therefore, the finite strain ε_{ii} and Poisson's ratio v_{ij} models for the twofold symmetrical tension and compression, as well as the more realistic compressive tetrachiral and anti-tetrachiral modes of mechanism, can be consolidated as

$$\varepsilon_{ii} = \ln$$

$$\frac{\left[\sqrt{2+\sqrt{2}}\cos\Delta\theta - \sqrt{2-\sqrt{2}}(\sin\Delta\theta + \cos\Delta\phi - \sin\Delta\phi)\right]\cos\Delta\Omega + 2l_i/l[g(l/l_i,\Delta\Omega)]}{\sqrt{2+\sqrt{2}} - \sqrt{2-\sqrt{2}} + 2l_i/l} \tag{4.14}$$

and

$$v_{ij} = -\frac{\ln\dfrac{\left[\sqrt{2+\sqrt{2}}\cos\Delta\theta - \sqrt{2-\sqrt{2}}(\sin\Delta\theta + \cos\Delta\phi - \sin\Delta\phi)\right]\cos\Delta\Omega + 2l_j/l[g(l/l_j,\Delta\Omega)]}{\sqrt{2+\sqrt{2}} - \sqrt{2-\sqrt{2}} + 2l_j/l}}{\ln\dfrac{\left[\sqrt{2+\sqrt{2}}\cos\Delta\theta - \sqrt{2-\sqrt{2}}(\sin\Delta\theta + \cos\Delta\phi - \sin\Delta\phi)\right]\cos\Delta\Omega + 2l_i/l[g(l/l_i,\Delta\Omega)]}{\sqrt{2+\sqrt{2}} - \sqrt{2-\sqrt{2}} + 2l_i/l}} \tag{4.15}$$

respectively, where

$$g(l/l_i,\Delta\Omega)$$

$$= \begin{cases} 1 & \Leftarrow \text{2-fold sym.} \\ \cos\left\{\sin^{-1}\left[0.5\left(\sqrt{2+\sqrt{2}} - \sqrt{2-\sqrt{2}}\right)l/l_i\sin\Delta\Omega\right]\right\} & \Leftarrow \text{tetrachiral} \\ 1 & \Leftarrow \text{anti-tetrach.} \end{cases} \tag{4.16}$$

and

$$\Delta\Omega = 0 \qquad \Leftarrow \quad \text{2-foldsym.}$$
$$\Delta\theta = \Delta\phi = 0 \quad \Leftarrow \quad \text{tetrachiral.} \qquad\qquad (4.17)$$
$$\Delta\theta = \Delta\phi = 0 \quad \Leftarrow \quad \text{anti-tetrach.}$$

As before, the Poisson's ratio described in Eq. (2.10) is recovered for finite deformation under square array regardless of the mode of mechanism.

The maximum rotation in the case of anti-tetrachiral mechanism can be sought by aligning the hinge rods parallel to the axes, as shown in Fig. 4.4, such that the rigid body rotation of the Maltese cross is arrested when the

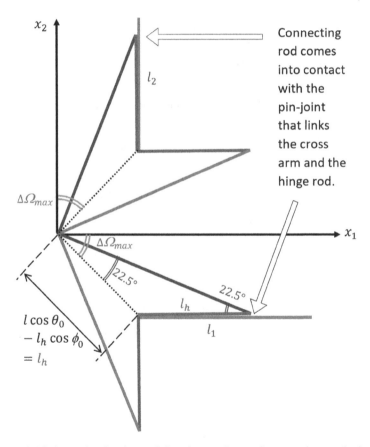

Figure 4.4 Schematics for determining the maximum free-rotation angle during anti-tetrachiral compression.

connecting rods come into contact with their nearest hinge rods. Since the hinge rods are originally aligned at an angle of $\pm45°$, the maximum rigid body rotation magnitude of the Maltese cross under anti-tetrachiral compression is $\Delta\Omega_{max} = 45°$. Unlike the case of the tetrachiral mechanism, the connecting rods parallel to both axes come into contact with their nearest hinge rods simultaneously.

4.4 Results

As an alternative to the tetrachiral mechanism, the anti-tetrachiral mechanism displayed in Fig. 2.7(c) has been developed to describe the deformation mechanism under the action of on-axis uniaxial load. Visualization of how the overall Poisson's ratio changes with geometry and the imposed compressive strain is shown in Fig. 4.5, whereby the infinitesimal v_{ij} (thick dashed lines) and finite v_{ij} (thin continuous curves) versus ε_{ii} were plotted for various $1 \leq 2l_i/l \leq 2$ at fixed $2l_j/l = 1.5$ (a) and various $1 \leq 2l_j/l \leq 2$ at fixed $2l_i/l = 1.5$ (b) as a result of anti-tetrachiral compression. The boundary for rigid-body rotation of the Maltese cross described is $d\Omega_{max} = 45°$ for anti-tetrachiral compression, and it is indicated in Fig. 4.5 as the boundary that connects the exact Poisson's ratio at the maximum compressive strain. Again, the results suggest that the geometrical properties (e.g. l, l_h, l_1, l_2) possess a more significant effect than the strain and the related rotation $d\Omega$.

5 Shear Mechanism

5.1 Trellis Shear Mechanism

With reference to the pin-jointed structure in its original states shown in Fig. 5.1 (top left), in which the equi-spaced rods are aligned to the axes described in Fig. 5.1 (bottom left), an application of pure shear with $\tau_{12} > 0$ transforms it to the shape shown in Fig. 5.1 (top right). This sheared shape can also be obtained upon application of tensile or compressive loads in the 45° and $-45°$ directions, respectively. Obviously, the transformed shape is reversed to that displayed in Fig. 5.1 (bottom right) upon reversal of the previously mentioned loads. The shear deformation illustrated in Fig. 2.7(d) approximates the Trellis shear if the size of the Maltese crosses diminishes. Since the shearing mode of the metamaterial is a modification of the Trellis shear, the term adopted for the deformation described in Fig. 2.7(d) is simply called the "shear" mechanism. Nevertheless, an understanding of the Trellis shear is useful, as it is adapted for the analysis of shear mechanism in this section. Reference to Fig. 5.1 also suggests that such a mechanism allows the metamaterial to

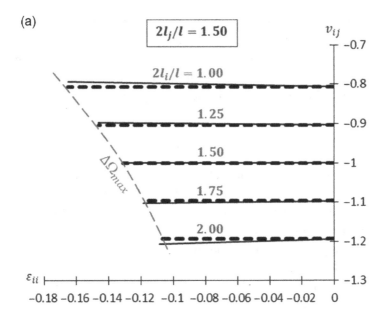

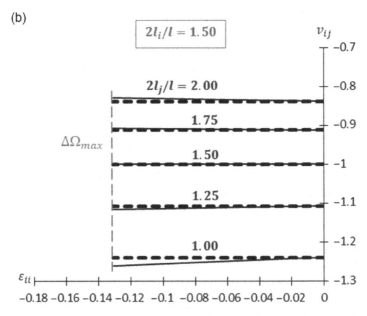

Figure 4.5 Effect of metamaterial spacing on Poisson's ratio v_{ij} under prescribed strain $\varepsilon_{ii} < 0$ for (a) various l_i at fixed l_j and (b) various l_j at fixed l_i based on the anti-tetrachiral mechanism. The dashed lines and continuous curves denote the infinitesimal and finite models, respectively, while the thin dashed lines indicate the maximum rotation defined in Fig. 4.4.

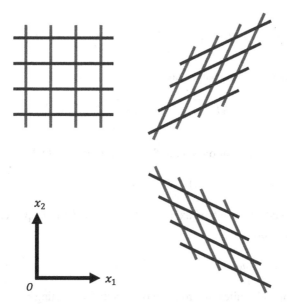

Figure 5.1 Schematics of trellis shearing.

function as a sieve that discriminates particle shapes rather than particle sizes, that is, it permits particles of spherical and regular polyhedral shape to pass through but reduces the number of long particles passing through. However, upon application of in-plane shearing or in-plane diagonal loads, the metamaterial sheet prevents polyhedral shapes from passing through but permits long particles to pass through.

5.2 Infinitesimal Deformation

The shearing mechanism identified in Fig. 2.7(d) is shown with greater detail in Fig. 5.2 for analysis in the case of square array, whereby the connecting rods originally aligned along the Ox_1 and Ox_2 axes rotate by an angle ω in the counterclockwise and clockwise directions, respectively, such that P, Q, and R displace to P', Q', and R', respectively, whereby the distance from the center of each Maltese cross to each inward pointed tip is

$$L = l \cos\frac{\pi}{8} - l_h \cos\frac{\pi}{4} = \frac{\sqrt{2+\sqrt{2}} - \sqrt{2-\sqrt{2}}}{2} l. \tag{5.1}$$

For a square array, the connecting rods of half-lengths l_1 and l_2 aligned along the Ox_1 and Ox_2, respectively, are common, namely l_c.

With reference to Fig. 5.2 (a),

$$\left(\overline{OP}\right) = \left(\overline{OR}\right) = 2L + 2l_c = \left(\sqrt{2 + \sqrt{2}} - \sqrt{2 - \sqrt{2}} \right) l + 2l_c \qquad (5.2)$$

so that the diagonal dimensions are

$$\left(\overline{OQ}\right) = \left(\overline{PR}\right) = \sqrt{2} \left(\sqrt{2 + \sqrt{2}} - \sqrt{2 - \sqrt{2}} \right) l + 2\sqrt{2} l_c. \qquad (5.3)$$

For an infinitesimal rotation of ω in the connecting rods, Q displaces to Q' by

$$d\left(\overline{OQ}\right)_i = -2l_c(1 - \cos \omega) + 2l_c \sin \omega \approx 2l_c \omega \qquad (5.4)$$

when evaluated along either axis ($i = 1, 2$), so that the displacement in the diagonal direction is

$$d\left(\overline{OQ}\right) = 2\sqrt{2} l_c \omega. \qquad (5.5)$$

Similarly, for an infinitesimal rotation of ω in the connecting rods, the distance PR shortens to P'R' by the amount

$$d\left(\overline{PR}\right)_i = -2l_c \sin \omega - 2l_c(1 - \cos \omega) \approx -2l_c \omega \qquad (5.6)$$

when projected along either axis ($i = 1, 2$), such that the actual change in length becomes

$$d\left(\overline{PR}\right) = -2\sqrt{2} l_c \omega. \qquad (5.7)$$

As such, we have the infinitesimal strains along the $Ox_{1'}$ and $Ox_{2'}$ axes, which are rotated by 45° counterclockwise from the Ox_1 and Ox_2 axes, as $\varepsilon_{1'1'} = d\left(\overline{OQ}\right)/\left(\overline{OQ}\right)$ and $\varepsilon_{2'2'} = d\left(\overline{PR}\right)/\left(\overline{PR}\right)$ or

$$\varepsilon_{1'1'} = -\varepsilon_{2'2'} = \frac{\omega}{\frac{\sqrt{2+\sqrt{2}} - \sqrt{2-\sqrt{2}}}{2} \frac{l}{l_c} + 1}. \qquad (5.8)$$

This mechanism as furnished in Fig. 5.2 (b) is not only attainable upon application of pure shear, as it can also be attained when a uniaxial tension or compression is applied along the ±45° direction. Considering the latter, we have the corresponding Poisson's ratio

$$\nu_{1'2'} = \nu_{2'1'} = 1. \qquad (5.9)$$

In other words, the Maltese cross system with square array exhibits the following in-plane Poisson's ratio

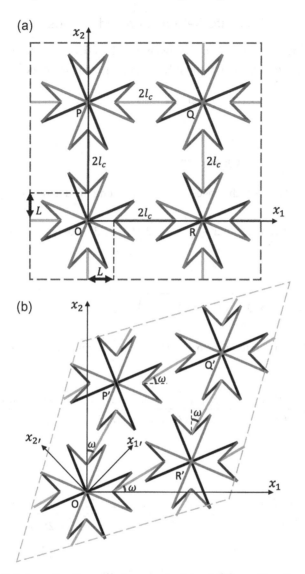

Figure 5.2 Schematics for pure shear analysis and diagonal loading, showing a 2-by-2 square array in original state (a) and upon shearing or $\pm45°$ loading (b).

$$v = \begin{cases} -1 \Leftrightarrow 0°, 90° \text{ loading} \\ +1 \Leftrightarrow \pm45° \text{ loading} \end{cases}$$

$$(5.10)$$

under infinitesimal deformation. The uniaxial off-axis loading of the present model can be viewed as a slight modification of the wine-rack model, which is auxetic in certain directions, as shown by Caruana-Gauci et al. [288].

5.3 Finite Deformation and Bounds

With reference to Fig. 5.2 (b) for obtaining the Poisson's ratio under finite rotation ω, the distances OQ$'$ and P$'$R$'$ when projected onto either axis give

$$\left(\overline{OQ'}\right)_i = 2L + 2l_c(\cos\omega + \sin\omega) \tag{5.11}$$

and

$$\left(\overline{P'R'}\right)_i = 2L + 2l_c(\cos\omega - \sin\omega) \tag{5.12}$$

for $i = 1, 2$. The actual distances as measured along the diagonals are therefore

$$\left(\overline{OQ'}\right) = \sqrt{2}\left(\sqrt{2+\sqrt{2}} - \sqrt{2-\sqrt{2}}\right)l + 2\sqrt{2}l_c(\cos\omega + \sin\omega) \tag{5.13}$$

and

$$\left(\overline{P'R'}\right) = \sqrt{2}\left(\sqrt{2+\sqrt{2}} - \sqrt{2-\sqrt{2}}\right)l + 2\sqrt{2}l_c(\cos\omega - \sin\omega) \tag{5.14}$$

respectively. With reference to Eq. (5.3), one obtains the corresponding true strains

$$\varepsilon_{1'1'} = \ln\frac{\left(\overline{OQ'}\right)}{\left(\overline{OQ}\right)} = \ln\frac{\sqrt{2+\sqrt{2}} - \sqrt{2-\sqrt{2}} + 2l_c/l(\cos\omega + \sin\omega)}{\sqrt{2+\sqrt{2}} - \sqrt{2-\sqrt{2}} + 2l_c/l} \tag{5.15}$$

and

$$\varepsilon_{2'2'} = \ln\frac{\left(\overline{P'R'}\right)}{\left(\overline{PR}\right)} = \ln\frac{\sqrt{2+\sqrt{2}} - \sqrt{2-\sqrt{2}} + 2l_c/l(\cos\omega - \sin\omega)}{\sqrt{2+\sqrt{2}} - \sqrt{2-\sqrt{2}} + 2l_c/l} \tag{5.16}$$

along the Ox_1' and Ox_2' axes, respectively. Therefore,

$$\nu_{1'2'} = \frac{1}{\nu_{2'1'}} = -\frac{\varepsilon_{2'2'}}{\varepsilon_{1'1'}} = -\frac{\ln\frac{\sqrt{2+\sqrt{2}}-\sqrt{2-\sqrt{2}}+2l_c/l(\cos\omega-\sin\omega)}{\sqrt{2+\sqrt{2}}-\sqrt{2-\sqrt{2}}+2l_c/l}}{\ln\frac{\sqrt{2+\sqrt{2}}-\sqrt{2-\sqrt{2}}+2l_c/l(\cos\omega+\sin\omega)}{\sqrt{2+\sqrt{2}}-\sqrt{2-\sqrt{2}}+2l_c/l}}. \tag{5.17}$$

Perusal of Fig. 5.2 (b) indicates that the limits to the counterrotations of the connecting rods are set by the hinge rods, namely $-45° < \omega < 45°$. As a result, the Poisson's ratio $\nu_{1'2'}$ is bounded within

$$-\frac{\ln\frac{\sqrt{2+\sqrt{2}}-\sqrt{2-\sqrt{2}}+2\sqrt{2}l_c/l}{\sqrt{2+\sqrt{2}}-\sqrt{2-\sqrt{2}}+2l_c/l}}{\ln\frac{\sqrt{2+\sqrt{2}}-\sqrt{2-\sqrt{2}}}{\sqrt{2+\sqrt{2}}-\sqrt{2-\sqrt{2}}+2l_c/l}} < v_{1'2'} < -\frac{\ln\frac{\sqrt{2+\sqrt{2}}-\sqrt{2-\sqrt{2}}}{\sqrt{2+\sqrt{2}}-\sqrt{2-\sqrt{2}}+2l_c/l}}{\ln\frac{\sqrt{2+\sqrt{2}}-\sqrt{2-\sqrt{2}}+2\sqrt{2}l_c/l}{\sqrt{2+\sqrt{2}}-\sqrt{2-\sqrt{2}}+2l_c/l}}$$

$$(5.18)$$

or, more conveniently,

$$-\frac{\ln\frac{1.0824+2.828l_c/l}{1.0824+2l_c/l}}{\ln\frac{1.0824}{1.0824+2l_c/l}} < v_{1'2'} < -\frac{\ln\frac{1.0824}{1.0824+2l_c/l}}{\ln\frac{1.0824+2.828l_c/l}{1.0824+2l_c/l}}. \qquad (5.19)$$

Since the denominator and the numerator of the lower and upper bounds, respectively, are negative, it follows that $v_{1/2'}$ is positive.

5.4 Results

A family of curves for the Poisson's ratio $v_{1/2'}$ for applied load along the Ox_1' axis versus the connecting rod rotation ω is plotted in Fig. 5.3(a) based on Eq. (5.17) at various $2l_c/l$ ratios for $\omega \neq 0$, in which the limits of the range $-45° < \omega < 45°$ are set by the hinge rods. Since Eq. (5.17) is undefined for $\omega = 0$, the Poisson's ratio $v_{1/2'} = 1$ based on the infinitesimal model is used therein. To display the validity, or the lack thereof, of the infinitesimal Poisson's ratio, the line $v_{1/2'} = 1$ is added; it can be seen that the infinitesimal model is correct only for $\omega = 0$. Equations (5.15) and (5.17) were used for plotting the Poisson's ratio curves against the prescribed strain along the Ox_1' axis shown in Fig. 5.3(b). Obtained results suggest that the prescribed strain plays a major role during compression, but the metamaterial unit spacing plays an increasing role during tension.

6 Deformation Paths and Poisson's Ratio (Dis)continuity

6.1 Deformation Pathways and Validity of Infinitesimal Models

The deformation path of interest is that pertaining to the Poisson's ratio, namely how the Poisson's ratio of this metamaterial evolves with the prescribed strain. This is visually represented in Fig. 6.1, in which the twofold symmetrical mechanism occurs, arising from applied on-axis uniaxial tension, but is least likely to arise from applied on-axis uniaxial compression. If one considers the stiffness of the twofold symmetrical mechanism and the group of tetrachiral and anti-tetrachiral mechanisms during on-axis compression, the former is least likely to occur since it has by far the highest stiffness of the three because of the central springs, which give it a relatively high stiffness in comparison to the

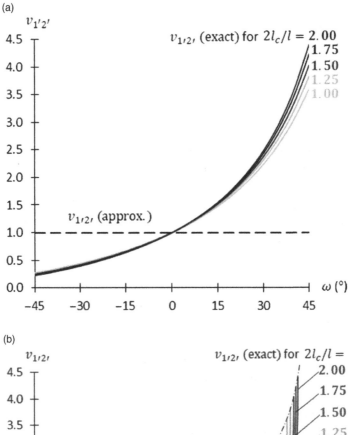

Figure 5.3 A family of $v_{1/2'}$ curves plotted against (a) connecting rod rotation ω, and (b) strain in the diagonal direction for various $2l_c/l$ ratios.

other two. For the twofold symmetrical mechanism to arise from applied on-axis uniaxial compression, a restrictive condition must be imposed whereby the Maltese crosses are not permitted to rotate. This imposed restriction is impractical, which is why the twofold symmetrical mechanism during applied on-axis uniaxial compression has been indicated as being highly idealized. The group of tetrachiral and anti-tetrachiral mechanisms are both driven by deformation of the "zero"-stiffness hinges. A proposal to determine whether the tetrachiral mechanism or the anti-tetrachiral mechanism will occur during on-axis loading is furnished in Section 7.2, "Open Problems and Recommendations for Future Work."

The possibility for the metamaterial to deform via contrasting mechanisms to exhibit Poisson's ratio of both signs suggests that it can be used as a conventional material when positive Poisson's ratio is required and as an auxetic material when negative Poisson's ratio is required. Positive Poisson's ratio is advantageous when a material sheet is used for wrapping a solid with anticlastic surface while negative Poisson's ratio is advantageous when a material sheet is used for wrapping around a solid with synclastic surface. It is also known that in the case of fiber-reinforced composites, it is important to prevent fiber removal from the surrounding matrix material. Depending on the mode of fiber removal, the choice of Poisson's ratio sign can help to resist or facilitate fiber removal. One of the fiber composite failures is fiber pullout from the matrix material. Under tensile action along the axis of a fiber, a positive Poisson's ratio is disadvantageous due to the radial contraction. When the fiber is auxetic, the application of axial tensile stress on the fiber tends to radially expand the fiber such that this lateral expansion acts as a self-locking mechanism to prevent fiber pullout. The converse is true for fiber pushout, whereby the application of compressive stress along the axis of a fiber produces radial expansion if the fiber possesses positive Poisson's ratio. Again, this lateral expansion acts as a form of self-locking mechanism against fiber pushout. If the fiber material possesses negative Poisson's ratio, the application of compressive stress along the axis of the fiber causes radial shrinkage, which is not helpful for preventing fiber pushout. Reference to Fig. 6.1 suggests that the sign of the Poisson's ratio can be predetermined by rotating the metamaterial such that the loading direction brings forth the desired Poisson's ratio sign.

The presence of logarithmic terms in the Poisson's ratio descriptions of the metamaterial for finite deformation does not permit them to be made concise as in the case of infinitesimal deformation. The simplicity of the infinitesimal models would naturally incentivize their adoption over the finite models. It therefore follows that there is a need to evaluate the validity of applying the infinitesimal models for finite deformation. Perusal of Fig. 2.6, Fig. 3.5, and

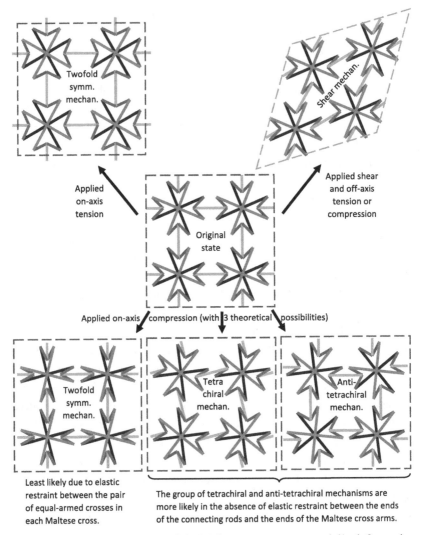

Figure 6.1 A visual summary of the Maltese cross metamaterial's deformation pathways in response to the various applied loads.

Fig. 4.5 suggests that use of the infinitesimal models is valid for finite deformation in the case of the twofold symmetrical mechanism, tetrachiral mechanism, and anti-tetrachiral mechanism.

6.2 Limitations

Despite the validity of the simpler infinitesimal models for finite deformation of the twofold symmetrical mechanism, tetrachiral mechanism, and anti-tetrachiral mechanism, the same does not apply for shear deformation as well as uniaxial loadings in the diagonal directions. It can be observed in Fig. 5.3 that the Poisson's ratios for shear and diagonal loadings according to the infinitesimal and finite models are in agreement only at $\varepsilon_{ii} = 0$. The first derivatives $\partial v_{ij}/\partial \varepsilon_{ii}$ are distinct for both models at zero prescribed strain. For this reason, the Poisson's ratio description for shear and diagonal uniaxial loadings based on infinitesimal deformation is not valid for finite deformation. Table 6.1 summarizes the Poisson's ratio models for various loading modes under infinitesimal and finite deformations, accompanied by remarks on the validity of the former's models for the latter's conditions.

6.3 Poisson's Ratio Continuity and Discontinuity

Arising from the greater likelihood for the metamaterial to deform in the tetrachiral or anti-tetrachiral manner rather than the symmetrically idealized mechanism during applied uniaxial compression, Fig. 6.2 combines the Poisson's ratio based on uniaxial tension and compression under different modes of mechanism.

It can be observed that (a) there is a Poisson's ratio discontinuity at zero strain between the twofold symmetrical mechanism during stretching and the tetrachiral mechanism during compression, as shown in Fig. 6.2 (top), but (b) no discontinuity at zero strain between the twofold symmetrical mechanism during stretching and the anti-tetrachiral mechanism during compression, as shown in Fig. 6.2 (bottom). For the latter, the continuity of Poisson's ratio at zero strain implies that

$$\lim_{\varepsilon_{ii} \to 0^-} v_{ij} = \lim_{\varepsilon_{ii} \to 0^+} v_{ij} = v_{ij}. \tag{6.1}$$

This means that for (b), a Poisson's ratio can be stated as a material property without the need to specify whether the applied load is tensile or compressive. For the former, the discontinuity of Poisson's ratio at zero strain reveals that

$$\lim_{\varepsilon_{ii} \to 0^-} v_{ij} \neq \lim_{\varepsilon_{ii} \to 0^-} v_{ij}. \tag{6.2}$$

This means that for (a), there are two Poisson's ratio curves, one each for tension and compression. There is therefore a need to specify whether the applied load is tensile or compressive, even if the applied load is infinitesimal.

The continuity discrepancy may be attributed to the motion of the connecting rods. For the twofold symmetrical mechanism and the anti-tetrachiral mechanism, all the connecting rods that are oriented on the same line remain oriented within a line even as the connecting rods (and hence the line) move. For the tetrachiral mechanism, all the connecting rods that are aligned on the same line

Table 6.1 Summary and validity remarks on Poisson's ratio models

Loading modes and deformation mechanisms	Infinitesimal models	Finite models	Remarks
Twofold symmetrical mechanism due to on-axis stretching	$v_{ij} = -\dfrac{k+2l_i/l}{k+2l_j/l}$	$v_{ij} = -\dfrac{\ln\dfrac{k_1\cos\Delta\theta - k_2(\sin\Delta\theta + \cos\Delta\phi - \sin\Delta\phi) + 2l_j/l}{k+2l_j/l}}{\ln\dfrac{k_1\cos\Delta\theta - k_2(\sin\Delta\theta + \cos\Delta\phi - \sin\Delta\phi) + 2l_i/l}{k+2l_i/l}}$	Infinitesimal model is valid for finite deformation.
Tetrachiral mechanism due to on-axis compression	$v_{ij} = -\dfrac{l_i}{l_j}$	$v_{ij} = -\dfrac{\ln\dfrac{k\cos\Delta\Omega + 2l_j/l\cos\left\{\sin^{-1}\left[\frac{k}{2}\frac{l}{l_j}\sin\Delta\Omega\right]\right\}}{k+2l_j/l}}{\ln\dfrac{k\cos\Delta\Omega + 2l_i/l\cos\left\{\sin^{-1}\left[\frac{k}{2}\frac{l}{l_i}\sin\Delta\Omega\right]\right\}}{k+2l_i/l}}$	Infinitesimal model is valid for finite deformation.
Ani-tetrachiral mechanism due to on-axis compression	$v_{ij} = -\dfrac{k+2l_i/l}{k+2l_j/l}$	$v_{ij} = -\dfrac{\ln\dfrac{k\cos\Delta\Omega + 2l_j/l}{k+2l_j/l}}{\ln\dfrac{k\cos\Delta\Omega + 2l_i/l}{k+2l_i/l}}$	Infinitesimal model is valid for finite deformation.
Shear mechanism due to shear and diagonal loadings	$v_{12} = 1$	$v_{12} = -\dfrac{\ln\dfrac{k+2l_c/l(\cos\omega - \sin\omega)}{k+2l_c/l}}{\ln\dfrac{k+2l_c/l(\cos\omega + \sin\omega)}{k+2l_c/l}}$	Infinitesimal model is not valid for finite deformation.

Note: The following constants are used for brevity, $k_1 = \sqrt{2+\sqrt{2}}$, $k_2 = \sqrt{2-\sqrt{2}}$, and $k = \sqrt{2+\sqrt{2}} - \sqrt{2-\sqrt{2}}$.

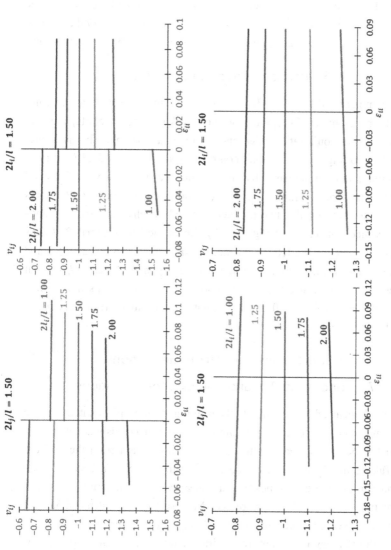

Figure 6.2 Effect of metamaterial spacing on the exact Poisson's ratio v_{ij} under prescribed strain ε_{ii} for various l_i at fixed l_j (left) and various l_j at fixed l_i (right) based on the twofold symmetrical mechanism under prescribed $\varepsilon_{ii} > 0$, as well as the tetrachiral mechanism (top) and anti-tetrachiral mechanism (bottom) under prescribed $\varepsilon_{ii} < 0$.

rotate by the same angle. Although the connecting rods that are originally on the same line are still parallel to each other after rotation, their axes no longer pass through the same line. In other words, the total change in the metamaterial size arising from the twofold symmetrical mechanism and the anti-tetrachiral mechanism are only due to the change in shape of the Maltese crosses and the rotation of the Maltese crosses, respectively, while the total change in the size of the metamaterial arising from the tetrachiral mechanism is due to *both* the rotation of the Maltese crosses *and* an opposite rotational motion of the connecting rods.

6.4 Poisson's Ratio Contour Maps

Having demonstrated that the effect of applied strain is insignificant in comparison to the geometrical properties, we next look into the effect of l_i and l_j on the effective Poisson's ratio under on-axis uniaxial loading. This is implemented by plotting v_{ij} for various combinations of $2l_i/l$ and $2l_j/l$ ratio at infinitesimal strain on the basis that the infinitesimal model is valid for small to moderate prescribed strain. Figure 6.3 (a) shows the infinitesimal Poisson's ratio contour plot for the twofold symmetrical mechanism and the anti-tetrachiral mechanism, while Fig. 6.3 (b) shows the infinitesimal Poisson's ratio contour plot for the tetrachiral mechanism for various $1 \le 2l_i/l \le 2$ and $1 \le 2l_j/l \le 2$. The former exhibits a more negative v_{ij} than the latter for $l_i < l_j$ but becomes less negative when $l_i > l_j$. In both cases, the Poisson's ratio is less negative, that is, $-1 < v_{ij} < 0$, for $l_i < l_j$, but becomes more negative, namely $v_{ij} < -1$, for $l_i > l_j$.

7 Conclusions and Recommendations

7.1 Concluding Remarks and Current Limitations

The technological advancement of 3D printing permits pin-jointed linkage mechanisms to be developed. A pin-jointed, linkage-mechanism auxetic metamaterial has been explored herein by drawing inspiration from the Maltese cross. Taking the Maltese cross geometry as the undeformed state of the metamaterial, prescription of positive and negative uniaxial strains in the on-axis direction causes the four arms of each Maltese cross to widen and narrow, respectively. Since this mechanism is less likely to occur during on-axis compression, the tetrachiral and anti-tetrachiral modes of mechanism permit in-plane contraction via rigid-body rotation of the Maltese crosses. Suppose a group of 2-by-2 metamaterial units are made to have the propensity to deform according to the tetrachiral and the anti-tetrachiral mechanisms; then, based on geometrical compatibility, the boundary changes of these four units permeate throughout the entire metamaterial structure so as to attain total tetrachiral or

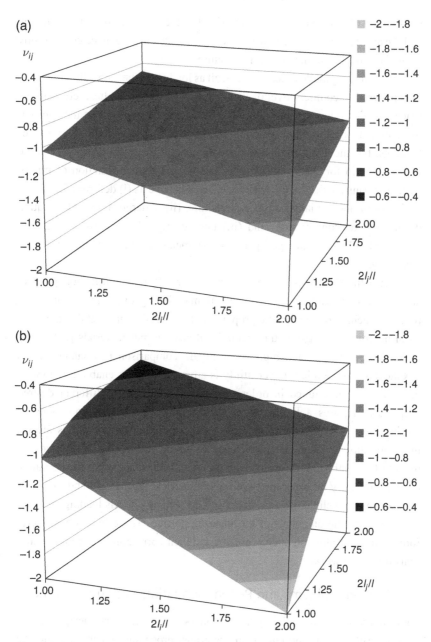

Figure 6.3 Contour plots of infinitesimal v_{ij} at various $2l_i/l$ and $2l_j/l$ ratios for the twofold symmetrical and anti-tetrachiral mechanisms (a) as well as the tetrachiral mechanism (b).

anti-tetrachiral mechanisms, respectively. In other words, both the tetrachiral and the anti-tetrachiral mechanisms are not expected to take place concurrently throughout the same metamaterial structure.

Under shearing mode of loading as well as loading in the $\pm 45°$ directions, the metamaterial encounters overall distortion whereby the Maltese crosses displace as rigid bodies without rotation. Plotted results suggest that the approximate model is valid for predicting the overall Poisson's ratio for on-axes loadings, but the exact models are recommended for quantifying the overall Poisson's ratio for shear and off-axes loadings. The manifestation of microstructural trinity – that is, the Maltese crosses behave as (i) deformable entities during prescribed uniaxial on-axis stretching, (ii) undergo rigid-body rotation during on-axis compression, and (iii) undergo rigid-body translation during shear and off-axis-loading – permits the metamaterial to manifest partial auxeticity.

Because the Maltese cross metamaterial is capable of exhibiting both positive and negative Poisson's ratio, as well as manifesting different mechanisms' routes for achieving the auxetic properties, it is therefore apt that this metamaterial be compared against the hierarchical auxetic metamaterials [289–301], which have demonstrated high versatility in comparison to nonhierarchical auxetic systems. While this versatility is advantageous for enabling the metamaterial to be fine-tuned in order to manifest significantly different desired properties, it poses a limitation for determining the deformation mode of mechanism where ambiguity exists. The main predictive limitations are (a) the ambiguity between the occurrence of tetrachiral versus anti-tetrachiral modes of mechanism if the propensity for either occurrence is not incorporated, (b) the ambiguity between the twofold symmetrical mechanism and the shear mechanism for a slight off-axis stretching, as well as off-axis loadings that are not parallel to the diagonals of the shearing portion. These limitations, therefore, guide the following open problems and their corresponding recommendations for future work.

7.2 Open Problems and Recommendations for Future Work

Some matters of practical importance on the microstructural design of this metamaterial include the choice of materials, processing technique, and the smallest unit size possible. An estimation of the size of each unit of the metamaterial and the choice of materials for constructing the linkages can be made based on current state-of-the-art technology. This can be established via reference to realistic nanoparticle sizes that have been successfully performed in 3D printing. If one were to use aluminum oxide (Al_2O_3) in 3D printing, the current Al_2O_3 nanoparticles can go down to the size of 50 nm, as reported by Wu et al. [302]. On the other hand, the adoption of iron oxide (Fe_3O_4) reveals that its nanoparticles can go down to the size of 15 nm, as recently reported by Domènech et al. [303]. Based on these data, one may assume a typical nanoparticle size of about 30 nm if these particles for 3D printing are made from metal oxides.

The smallest possible unit size can be estimated as follows. Let the thickness of the spiral spring be one order higher than the nanoparticle size, for example, $h = 300$ nm, then the diameter of the pin-joint can be set at another order higher, namely 3 μm, which would then allow the rod width to be 6 μm. Suppose the lengths of both the cross arm l and the hinge rod l_h are set at an order greater than their width of 6 μm; then, the choice of $l = 80$ μm and $l_h = 43.3$ μm would comply with Eq. (2.2). Imposing a square array such that $l_i = l = 80$ μm $(i = 1, 2)$ would then give a unit cell size of $2x_i = 247$ μm [281], thereby suggesting that the use of nanoparticle size in the order of 10^1 nm can attain a metamaterial unit of length scale 10^2 μm. On the other hand, the size of each unit should be enlarged if there is a need to produce a material of sizeable product. The maximum size of the entire metamaterial structure is limited by the printing dimension of the 3D printer, which can range from 500 mm × 280 mm × 360 mm   up   to   1800 mm × 1000 mm × 700 mm, depending on the specific 3D printer used. A survey on the smallest possible nanoparticles that can provide reasonably good structural integrity is suggested for future work to pave the way for construction of and experimentation with this metamaterial.

In practice, loading conditions often have a level of offset. The variation of the overall Poisson's ratio with the offset angle, addressing specific applications (such as indentation), as well as the implementation of spiral springs to all other hinges and the effective elastic moduli need to be done for future work. It is further suggested that experimentation and/or finite element analysis be performed on the Maltese cross metamaterial not only to validate the analytical models but also to evaluate the ratio of the various deformation modes, because the various mechanisms are **not** equally likely to occur. This is to quantify the relative occurrence of the various mechanisms, which may take place concurrently but to a different extent, if spiral springs are also attached at the joints between the connecting rods and the hinge rods.

A possible way to stabilize the unit cell during on-axis compression is to impose a level of stiffness at the joints between the connecting rod and the pair of hinge rods. Since there are three rods involved per pin-joint, that is, each end of a connecting rod is connected to a pair of hinge rods while a spiral spring has only two ends, the manner in which rotational stiffness is incorporated therein has to be different from the type of rotational stiffness imposed between the pair of equal-armed crosses. One possible way to apply rotational stiffness there is to impose sideward compression from the pair of hinge rods to the connecting rod by means of roller pairs, as illustrated in Fig. 7.1. As the name suggests, each of these roller pairs comprises a pair of assemblies consisting of springs, rods, and rollers that compresses the connecting rod on both sides with equal force so that the horizontal and vertical connecting rods remain horizontal and vertical, respectively, during a stress-free state.

An alternative is to introduce a pair of secondary spiral springs at each of the joints indicated as I, J, M, and N in Fig. 7.2. To ensure symmetry at the midplane of the metamaterial a set of spiral springs is attached at the upper surface (e.g. between hinge rod AJ and the connecting rod) and another set of springs is attached at the lower surface (e.g. between hinge rod EF and the connecting rod), such that during any relative rotation one set of spiral springs experience increasing curvature and another set decreasing curvature. These springs require very small constants sufficient to ensure that the connecting rods remain oriented to both axes during on-axis stretching but permit the Maltese crosses to rotate with respect to the connecting rods during on-axis compression, off-axis loading, and shear loading. The predictive limitations mentioned in Section 7.1 can be addressed by incorporating a level of rotational stiffness between the

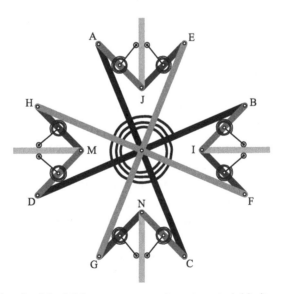

Figure 7.1 A unit of the Maltese cross auxetic metamaterial in its original state, incorporating rotational stiffness to the connecting rods by means of roller pairs at the hinge rods.

hinge rods and the connecting rods. With the presence of the primary and secondary springs, as illustrated in Fig. 7.1 and Fig. 7.2, there are thus combined modes of deformation.

Recall the first predictive limitation arising from the ambiguity between the occurrence of tetrachiral versus anti-tetrachiral modes of mechanism if the propensity for either occurrence is not incorporated. With the availability of primary and secondary spiral springs, two simultaneous modes of mechanism are identified for on-axis compression – the rotation of the Maltese crosses as shown in Fig. 2.7 (b) or (c), and the change in shape of the Maltese cross as shown in Fig. 2.2 (bottom). The former is influenced by the added rotational stiffness at the joint between the connecting and hinge rods, and the latter governed by the rotational stiffness at the joint between the pair of equal-armed crosses. For a given l_i/l_j ratio and rotational stiffness ratio of the main spring to the secondary spring, the determination of

- whether on-axis compression leads to tetrachiral mechanism or anti-tetrachiral mechanism,
- extent of Maltese cross rotation $\Delta\Omega$ vis-à-vis the change in Maltese cross shape as measured by either $\Delta\theta$ or $\Delta\phi$ during on-axis compression, and
- on-axis Young's modulus

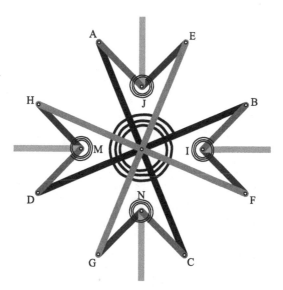

Figure 7.2 A unit of the Maltese cross auxetic metamaterial in its original state, incorporating rotational stiffness to the connecting rods by means of secondary spiral springs at joints I, J, M, and N. For clarity, only the upper level of secondary springs is shown.

can then be established by energy consideration. With reference to the second predictive limitation due to the ambiguity between the twofold symmetrical mechanism and the shear mechanism for a slight off-axis stretching, the metamaterial encounters both rearrangement of the Maltese crosses that resemble the shearing mechanism indicated in Fig. 2.7 (d) as well as the change in shape of the Maltese cross illustrated in Fig. 2.2 (top). Hence for a given l_i/l_j ratio, rotational stiffness ratio of the main spring to the secondary spring, and direction of stretching, the quantitative description of

- global rotation of the metamaterial,
- combined shearing mechanism and the twofold symmetrical mechanism, and
- Young's modulus in the direction of stretching

can again be sought by energy minimization. In the case of applied pure shearing, the shear modulus G_{12} is primarily controlled by the stiffness of the secondary spiral springs since the shearing mechanism is facilitated by the opposite rotational directions of the connecting rods in different alignment.

References

1. Pendry JB (2000), Negative refraction makes a perfect lens, *Phys. Rev. Lett.*, **85**(18), 3966–3969.
2. Caloz C and Itoh T (2006), *Electromagnetic Metamaterials: Transmission Line Theory and Microwave Applications.* Wiley, Hoboken, NJ.
3. Cui TJ, Smith DR, and Liu R (2010), *Metamaterials. Springer*, Boston, MA.
4. Grimberg R (2013), Electromagnetic metamaterials, *Mater. Sci. Eng. B*, **178**(19), 1285–1295.
5. Cui TJ, Tang WX, Yang XM, Mei ZL, and Jiang WX (2016), *Metamaterials: Beyond Crystals, Noncrystals, and Quasicrystals. CRC Press*, Boca Raton, FL.
6. Sakoda K (2019), *Electromagnetic Metamaterials.* Springer Nature, Singapore.
7. Fang N, Xi D, Xu J, Ambati M, Srituravanich W, Sun C, and Zhang X (2006), Ultrasonic metamaterials with negative modulus, *Nat. Mater.*, **5**(6), 452–456.
8. Pendry JB and Li J (2008), An acoustic metafluid: Realizing a broadband acoustic cloak, *New J. Phys.*, **10**(11), 115032.
9. Craster RV and Guenneau S (2013), *Acoustic Metamaterials.* Springer, Dordrecht.
10. Milton GW, Briane M, and Willis JR (2006), On cloaking for elasticity and physical equations with a transformation invariant form, *New J. Phys.*, **8** (10), 248.
11. Frenzel T, Kadic M, and Wegener M (2017), Three-dimensional mechanical metamaterials with a twist, *Science*, **358**(6366), 1072–1074.
12. Sujardi JU, Gao L, Du H, Li X, Xiong X, Fang NX, and Lu Y (2019), Mechanical metamaterials and their engineering applications, *Adv. Eng. Mater.*, **21**(3), 1800864.
13. Huidobro PA, Fernandez-Dominguez AI, Pendry JB, Martin-Moreno L, and Garcia-Vidal FJ (2018), *Spoof Surface Plasmon Metamaterials.* Cambridge University Press, Cambridge.
14. Li R, Wang Z, and Chen H (2020), *Metamaterials and Negative Refraction.* Cambridge University Press, Cambridge.
15. Cui TJ and Liu S (2021), *Information Metamaterials.* Cambridge University Press, Cambridge.
16. Jiang WX, Mei ZL, and Cui TJ (2021), *Effective Medium Theory of Metamaterials and Metasurfaces.* Cambridge University Press, Cambridge.
17. Li Y, Zhou Z, He Y, and Li H (2021), *Epsilon-Near-Zero Metamaterials.* Cambridge University Press, Cambridge.

18. Hummel FA (1951), Thermal expansion properties of some synthetic lithia minerals, *J. Am. Ceram. Soc.*, **34**(8), 235–239.

19. Chu CN, Saka N, and Suh NP (1987), Negative thermal expansion ceramics: A review, *Mater. Sci. Eng.*, **95**, 303–308.

20. Sleight AW (1995), Thermal contraction, *Endeavour*, **19**(2), 64–68.

21. Sleight AW (1998), Compounds that contract on heating, *Inorg. Chem.*, **37** (12), 2854–2860.

22. Sleight AW (1998), Isotropic negative thermal expansion, *Ann. Rev. Mater. Sci.*, **28**, 29–43.

23. Mary TA, Evans JSO, Vogt T, and Sleight AW (1996), Negative thermal expansion from 0.3 to 1050 Kelvins in ZrW2O8, *Science*, **272**(5258), 90–92.

24. Dove MT, Welche PRL, and Heine V (1998), Negative thermal expansion in beta-quartz, *Phys. Chem. Minerals*, **26**(1), 63–77.

25. Evans JSO, Hanson JC, and Sleight AW (1998), Room-temperature super-structure of ZrV2O7, *Acta Crystallog. B*, **54**(6), 705–713.

26. Evans JSO, Jorgensen JD, Short S, David WIF, Ibberson RM, and Sleight AW (1999), Thermal expansion in the orthorhombic phase of ZrW_2O_8, *Phys. Rev. B*, **60**(21), 14643–14648.

27. Lim TC (2005), Anisotropic and negative thermal expansion behavior in a cellular microstructure, *J. Mater. Sci.*, **40**(12), 3275–3277.

28. Barrera GD, Bruno JAO, Barron THK, and Allan NL (2005), Negative thermal expansion, *J. Phys.: Condens. Matter*, **17**(4), R217–R252.

29. Miller W, Smith CW, Mackenzie DS, and Evans KE (2009), Negative thermal expansion: A review, *J. Mater. Sci.*, **44**(20), 5441–5451.

30. Lim TC (2012), Negative thermal expansion structures constructed from positive thermal expansion trusses, *J. Mater. Sci.*, **47**(1), 368–373.

31. Lim TC (2013), Negative thermal expansion in transversely isotropic space frame trusses, *Phys. Status Solidi B*, **250**(10), 2062–2069.

32. Lim TC (2017), Auxetic and negative thermal expansion structure based on interconnected array of rings and sliding rods, *Phys. Status Solidi B*, **254** (12), 1600775.

33. Attfield JP (2018), Mechanisms and materials for NTE, *Front. Chem.*, **6**, 371.

34. Ai L and Gao XL (2018), Three-dimensional metamaterials with a negative Poisson's ratio and a non-positive coefficient of thermal expansion, *Int. J. Mech. Sci.*, **135**, 101–113.

35. Takenaka K (2018), Progress of research in negative thermal expansion materials: Paradigm shift in the control of thermal expansion, *Front. Chem.*, **6**, 267.

36. Lim TC (2019), A 2D auxetikos system based on interconnected shurikens, *SN Appl. Sci.*, **1**(11), 1383.

37. Lim TC (2019), 2D metamaterial with in-plane positive and negative thermal expansion and thermal shearing based on interconnected alternating bimaterials, *Mater. Res. Exp.*, **6**(11), 115804.

38. Lim TC (2020), Metacomposite with auxetic and in situ sign reversible thermal expansivity upon temperature fluctuation, *Compos. Commun.*, **19**, 30–36.

39. Baughman RH, Stafstrom S, Cui C, and Dantas SO (1998), Materials with negative compressibilities in one or more dimensions, *Science*, **279**(5356), 1522–1524.

40. Gatt R and Grima JN (2008), Negative compressibility, *Phys. Status Solidi RRL*, **2**(5), 236–238.

41. Goodwin AL, Keen DA, and Tucker MG (2008), Large negative linear compressibility of $Ag_3[Co(CN)_6]$, *Proc. Nat. Acad. Sci.*, **105**(48), 18708–18713.

42. Lakes R and Wojciechowski KW (2008), Negative compressibility, negative Poisson's ratio, and stability, *Phys. Status Solidi B*, **245**(3), 545–551.

43. Grima JN, Attard D, and Gatt R (2008), Truss-type systems exhibiting negative compressibility, *Phys. Status Solidi B*, **245**(11), 2405–2414.

44. Fortes ADS, Suard E, and Knight KS (2011), Negative linear compressibility and massive anisotropic thermal expansion in methanol monohydrate, *Science*, **331**(6018), 742–746.

45. Grima JN, Attard D, Caruana-Gauci R, and Gatt R (2011), Negative linear compressibility of hexagonal honeycombs and related systems, *Scripta Mater.*, **65**(7), 565–568.

46. Grima JN, Caruana-Gauci R, Attard D, and Gatt R (2012), Three-dimensional cellular structures with negative Poisson's ratio and negative compressibility properties, *Proc. Royal Soc. A*, **468**(2146), 3121–3138.

47. Nicolaou ZG and Motter AE (2012), Mechanical metamaterials with negative compressibility transitions, *Nat. Mater.*, **11**(7), 608–613.

48. Cairns AB, Catafesta J, Levelut C, Rouquette J, van der Lee A, Peters L, Thompson AL, Dmitriev V, Haines J, and Goodwin AL (2013), Giant negative linear compressibility in zinc dicyanoaurate, *Nat. Mater.*, **12**(3), 212–216.

49. Grima JN, Caruana-Gauci R, Wojciechowski KW, and Evans KE (2013), Smart hexagonal truss systems exhibiting negative compressibility through constrained angle stretching, *Smart Mater. Struct.*, **22**(8), 084015.

50. Imre AR (2014), Metamaterials with negative compressibility – a novel concept with a long history, *Mater. Sci.-Poland*, **32**(2), 126–129.

51. Miller W, Evans KE, and Marmier A (2015), Negative linear compressibility in common materials, *Appl. Phys. Lett.*, **106**(23), 231903.

52. Cairns AB and Goodwin AL (2015), Negative linear compressibility, *Phys. Chem. Chem. Phys.*, **17**(32), 20449–20465.

53. Zhou X, Zhang L, Zhang H, Liu Q, and Ren T (2016), 3D cellular models with negative compressibility through the wine-rack-type mechanism, *Phys. Status Solidi B*, **253**(10), 1977–1993.

54. Attard D, Caruana-Gauci R, Gatt R, and Grima JN (2016), Negative linear compressibility from rotating rigid units, *Phys. Status Solidi B*, **253**(7), 1410–1418.

55. Dudek KK, Attard D, Caruana-Gauci R, Wojciechowski KW, and Grima JN (2016), Unimode metamaterials exhibiting negative linear compressibility and negative thermal expansion, *Smart Mater. Struct.*, **25**(2), 025009.

56. Lim TC (2017), 2D structures exhibiting negative area compressibility, *Phys. Status Solidi B*, **254**(12), 1600682.

57. Colmenero F (2019), Silver oxalate: mechanical properties and extreme negative mechanical phenomena, *Adv. Theory Simul.*, **2**(6), 1900040.

58. Degabriele EP, Attard D, Grima-Cornish JN, Caruana-Gauci R, Gatt R, Evans KE, and Grima JN (2019), On the compressibility properties of the wine-rack-like carbon allotropes and related poly(phenylacetylene), *Phys. Status Solidi B*, **256**(1), 1800572.

59. Grima-Cornish JN, Grima JN, and Attard D (2020), A novel mechanical metamaterial exhibiting auxetic behavior and negative compressibility, *Materials*, **13**(1), 79.

60. Barrett DG, Bushnell GG, and Messersmith PB (2013), Mechanically robust, negative-swelling, mussel-inspired tissue adhesives, *Adv. Healthcare Mater.*, **2**(5), 745–755.

61. Liu J, Gu T, Shan S, Kang SH, Weaver JC, and Bertoldi K (2016), Harnessing buckling to design architected materials that exhibit effective negative swelling, *Adv. Mater.*, **28**(31), 6619–6624.

62. Lim TC (2018), A negative hygroscopic expansion material, *Mater. Sci. Forum.*, **928**, 277–282.

63. Curatolo M (2018), Effective negative swelling of hydrogel-solid composites, *Extreme Mech. Lett.*, **25**, 46–52.

64. Zhang H, Gou X, Wu J, Fang D, and Zhang Y (2018), Soft mechanical metamaterials with unusual swelling behavior and tunable stress-strain curves, *Sci. Adv.*, **4** (6), eaar8535.

65. Lim TC (2019), A reinforced kite-shaped microstructure with negative linear and area hygrothermal expansions, *Key Eng. Mater.*, **803**, 272–277.

66. Lim TC (2019), Negative environmental expansion for interconnected array of rings and sliding rods, *Phys. Status Solidi B*, **256**(1), 1800032.

67. Lim TC (2020), Negative hygrothermal expansion of reinforced double arrowhead microstructure, *Phys. Status Solidi B*, **257**(10), 1800055.

68. Lakes RS (1993), Microbuckling instability in elastomeric cellular solids, *J. Mater. Sci.*, **28**(17), 4667–4672.

69. Wang YC and Lakes RS (2001), Extreme thermal expansion, piezoelectricity, and other coupled field properties in composites with a negative stiffness phase, *J. Appl. Phys.*, **90**(12), 6458–6465.

70. Lakes RS (2001), Extreme damping in compliant composites with a negative-stiffness phase, *Phil. Mag. Lett.*, **81**(2), 95–100.

71. Lakes RS (2001), Extreme damping in composite materials with a negative stiffness phase, *Phys. Rev. Lett.*, **86**(13), 2897–2900.

72. Lakes RS, Lee T, Bersie A, and Wang YC (2001), Extreme damping in composite materials with negative stiffness inclusions, *Nature*, **410**(6828), 565–567.

73. Lakes RS and Drugan WJ (2002), Dramatically stiffer elastic composite materials due to a negative stiffness phase?, *J. Mech. Phys. Solids*, **50**(5), 979–1009.

74. Wang YC and Lakes RS (2004), Extreme stiffness systems due to negative stiffness elements, *Am. J. Phys.*, **72**(1), 40–50.

75. Wang YC and Lakes RS (2004), Negative stiffness-induced extreme viscoelastic mechanical properties: Stability and dynamics, *Phil. Mag.*, **84**(35), 3785–3801.

76. Wang YC and Lakes RS (2005), Stability of negative stiffness viscoelastic systems, *Quart. Appl. Math.*, **63**(1), 34–55.

77. Wang YC, Swadener JG, and Lakes RS (2006), Two-dimensional viscoelastic discrete triangular system with negative-stiffness components, *Phil. Mag. Lett.*, **86**(2), 99–112.

78. Moore B, Jaglinski T, Stone DS, and Lakes RS (2006), Negative incremental bulk modulus in foams, *Phil. Mag. Lett.*, **86**(10), 651–659.

79. Dyskin AV and Pasternak E (2012), Mechanical effect of rotating non-spherical particles on failure in compression, *Phil. Mag.*, **92**(28–30), 3451–3473.

80. Rafsanjani A, Akbarzadeh A, and Pasini D (2015), Snapping mechanical metamaterials under tension, *Adv. Mater.*, **27**(39), 5931–5935.

81. Balch SP and Lakes RS (2017), Amelioration of waves and microvibrations by micro-buckling in open celled foam, *Cell. Polym.*, **36**(1), 1–12.

82. Che K, Yuan C, Wu J, Qi HJ, and Meaud J (2017), Three-dimensional-printed multistable mechanical metamaterials with a deterministic deformation sequence, *J. Appl. Mech.*, **84**(1), 011004.

83. Dudek KK, Gatt R, Dudek MR, and Grima JN (2018), Negative and positive stiffness in auxetic magneto-mechanical metamaterials, *Proc. Royal Soc. A*, **474**(2215), 20180003.

84. Ha CS, Lakes RS, and Plesha ME (2018), Design, fabrication, and analysis of lattice exhibiting absorption via snap through behaviour, *Mater. Des.*, **141**, 426–437.

85. Ha CS, Lakes RS, and Plesha ME (2019), Cubic negative stiffness lattice structure for energy absorption: Numerical and experimental studies, *Int. J. Solids Struct.*, **178–179**, 127–135.

86. Karachevtseva I, Pasternak E, and Dyskin AV (2019), Negative stiffness produced by rotation of non-spherical particles and its effect on frictional sliding, *Phys. Status Solidi B*, **256**(1), 1800003.

87. Pasternak E and Dyskin AV (2019), Architectured materials with inclusions having negative Poisson's ratio or negative stiffness. In Estrin Y, Brechet Y, Dunlop J, and Fratzl P (eds.), *Architectured Materials in Nature and Engineering*, pp. 51–87. Springer Nature, Cham.

88. Wang YC, Lai HW, and Shen MW (2019), Effects of cracks on anomalous mechanical behavior and energy dissipation of negative-stiffness plates, *Phys. Status Solidi B*, **256**(1), 1800489.

89. Zhang Y, Restrepo D, Velay-Lizancos M, Mankame ND, and Zavattieri PD (2019), Energy dissipation in functionally two-dimensional phase transforming cellular materials, *Scient. Rep.*, **9**, 12581.

90. Evans KE, Nkansah MA, Hutchinson IJ, and Rogers SC (1991), Molecular design network, *Nature*, **353**(6340), 124.

91. Voigt W (1928), *Lehrbuch der Kristallphysik*, Teubner Verlag, Leipzig.

92. Simmons G and Birch F (1963), Elastic constants of pyrite, *J. Appl. Phys.*, **34**(9), 2736–2738.

93. Benbattouche N, Saunders GA, Lambson EF, and Honle W (1989), The dependences of the elastic stiffness moduli and the Poisson ratio of natural iron pyrites FeS_2 upon pressure and temperature, *J. Phys. D. Appl. Phys.*, **22**(5), 670–675.

94. Li Y (1976), The anisotropic behavior of Poisson's ratio, Young's modulus, and shear modulus in hexagonal material, *Phys. Status Solidi*, **38**(1), 171–175.

95. Berlincourt D and Jaffe H (1958), Elastic and piezoelectric coefficients of single-crystal barium titanate, *Phys. Rev.*, **111**(1), 143–148.

96. Alderson A and Evans KE (2009), Deformation mechanisms leading to auxetic behaviour in the α-cristobalite and α-quartz structures of both silica and germania, *J. Phys. Condens. Matter*, **21**(2), 25401.

97. Kittinger E, Tichy J, and Bertagnolli E (1981), Example of a negative effective Poisson's ratio, *Phys. Rev. Lett.*, **47**(10), 712–714.

98. Milstein F and Huang K (1979), Existence of a negative Poisson ratio in fcc crystals, *Phys. Rev. B*, **19**(4), 2030–2033.

99. Baughman RH, Shacklette JM, Zakhidov AA, and Stafström S (1998), Negative Poisson's ratios as a common feature of cubic metals, *Nature*, **392**(6674), 362–365.

100. Ho DT, Park SD, Kwon SY, Park K, and Kim SY (2014), Negative Poisson's ratios in metal nanoplates, *Nat. Commun.*, **5**, 3255.

101. Gunton DJ and Saunders GA (1972), The Young's modulus and Poisson's ratio of arsenic, antimony and bismuth, *J. Mater. Sci.*, **7**(9), 1061–1068.

102. Yeganeh-Haeri A, Weidner DJ, and Parise JB (1992), Elasticity of α-cristobalite: A silicon dioxide with a negative Poisson's ratio, *Science*, **257**(5070), 650–652.

103. Homand-Etienne F and Houpert R (1989), Thermally induced microcracking in granites: Characterization and analysis, *Int. J. Rock Mech. Min. Sci.*, **26**(2), 125–134.

104. Peura M, Grotkopp I, Lemke H, Vikkula A, Laine J, Müller M, and Serima R (2006), Negative Poisson ratio of crystalline cellulose in kraft cooked Norway spruce, *Biomacromol.*, **7**(5), 1521–1528.

105. Nakamura K, Wada M, Kuga S, and Okano T (2004), Poisson's ratio of cellulose I_β and cellulose II, *J. Polym. Sci. Part B Polym. Phys.*, **42**(7), 1206–1211.

106. Yao YT, Alderson A, and Alderson KL (2012), Towards auxetic nanofibres: Molecular modelling of auxetic behaviour in cellulose II, *Behavior and Mechanics of Multifunctional Materials and Composites Proc. SPIE*, **8342**, 83421W.

107. Veronda DR and Westmann RA (1970), Mechanical characterization of skin-finite deformations, *J. Biomech.*, **3**(1), 111–124.

108. Lees C, Vincent JFV, and Hillerton JE (1991), Poisson's Ratio in skin, *Biomed. Mater. Eng.*, **1**(1), 19–23.

109. Song F, Zhou J, Xu X, Xu Y, and Bai Y (2008), Effect of a negative Poisson ratio in the tension of ceramics, *Phys. Rev. Lett.*, **100**(24), 245502.

110. Williams JL and Lewis JL (1982), Properties and an anisotropic model of cancellous bone from the proximal tibial epiphysis, *J. Biomech. Eng.*, **104**(1), 50–56.

111. Renson CE and Braden M (1975), Experimental determination of the rigidity modulus, Poisson's ratio and elastic limit in shear of human dentine, *Arch. Oral Biol.*, **20**(1), 43–47.

112. Schmidt CF, Svoboda K, Lei N, Petsche IB, Berman LE, Safinya CR, and Grest G (1993), Existence of a flat phase in red cell membrane skeletons, *Science*, **259**(5097), 952–955.

113. Chen X and Brodland GW (2009), Mechanical determinants of epithelium thickness in early-stage embryos, *J. Mech. Behav. Biomed. Mater.*, **2**(5), 494–501.

114. Timmins LH, Wu Q, Yeh AT, Moore JE, and Greenwald SE (2010), Structural inhomogeneity and fiber orientation in the inner arterial media, *Am. J. Physiol. Heart Circ. Physiol.*, **298**(5), H1537–H1545.

115. Patten K and Wess T (2013), Suprafibrillar structures of collagen, evidence for local organization and auxetic behaviour in architectures, *J. Biophys. Chem.*, **4**(3), 103–109.

116. Pagliara S, Franze K, McClain CR, Wylde GW, Fisher CL, Franklin RJM, Kabla AJ, Keyser UF, and Chalut KJ (2014), Auxetic nuclei in embryonic stem cells exiting pluripotency, *Nat. Mater.*, **13**(6), 638–644.

117. Tokmakova SP (2005), Stereographic projections of Poisson's ratio in auxetic crystals, *Phys. Status Solidi B*, **242**(3), 721–729.

118. Grima JN, Winczewski S, Mizzi L, Grech MC, Cauchi R, Gatt R, Attard D, Wojciechowski KW, and Rybicki J (2015), Tailoring graphene to achieve negative Poisson's ratio properties, *Adv. Mater.*, **27**(8), 1455–1459.

119. Aouni N and Wheeler L (2008), Auxeticity of calcite and aragonite polymorphs of $CaCO_3$ and crystals of similar structure, *Phys. Status Solidi B*, **245**(11), 2454–2462.

120. Jiang JW and Park HS (2014), Negative Poisson's ratio in single-layer black phosphorus, *Nat. Commun.*, **5**, 4727.

121. Jiang JW, Rabczuk T, and Park HS (2015), A Stillinger–Weber potential for single-layered black phosphorus, and the importance of cross-pucker interactions for a negative Poisson's ratio and edge stress-induced bending, *Nanoscale*, **7**(14), 6059–6068.

122. Jiang JW (2015), Thermal conduction in single-layer black phosphorus: Highly anisotropic?, *Nanotechnol.*, **26**(5), 055701.

123. Rovati M (2003), On the negative Poisson's ratio of an orthorhombic alloy, *Scr. Mater.*, **48**(3), 235–240.

124. Kellogg RA, Russell AM, Lograsso TA, Flatau AB, Clark AE, and Wun-Fogle M (2003), Mechanical properties of magnetostrictive iron-gallium alloys, *SPIE Smart Structures and Materials 2003, Active Materials: Behaviour and Mechanics, Proc.*, **5053**, 70.

125. Kellogg RA, Russell AM, Lograsso TA, Flatau AB, Clark AE, and Wun-Fogle M (2004), Tensile properties of magnetostrictive iron–gallium alloys, *Acta Mater.*, **52**(7), 5043–5050.

126. Petculescu G, Hathaway KB, Lograsso TA, Wun-Fogle M, and Clark AE (2005), Magnetic field dependence of galfenol elastic properties, *J. Appl. Phys.*, **97**(10), 10M315.

127. Schurter HM and Flatau AB (2008), Elastic properties and auxetic behavior of Galfenol for a range of compositions, *Behavior and Mechanics of Multifunctional and Composite Materials, Proc. SPIE*, **6929**, 69291U.

128. Valant M, Axelsson AK, Aguesse F, and Alford NM (2010), Molecular auxetic behavior of epitaxial co-ferrite spinel thin film, *Adv. Funct. Mater.*, **20**(4), 644–647.

129. Zhang Y, Wu R, Schurter HM, and Flatau AB (2010), Understanding of large auxetic properties of iron-gallium and iron-aluminum alloys, *J. Appl. Phys.*, **108**(2), 023513.

130. Li D, Jaglinski T, Stone DS, and Lakes RS (2012), Temperature insensitive negative Poisson's ratios in isotropic alloys near a morphotropic phase boundary, *Appl. Phys. Lett.*, **101**(25), 251903.

131. Wang XF, Jones TE, Li W, and Zhou YC (2012), Extreme Poisson's ratios and their electronic origin in B2 CsCl-type AB intermetallic compounds, *Phys. Rev. B*, **85** (13), 134108.

132. Lim TC (2007), On simultaneous positive and negative Poisson's ratio laminates, *Phys. Status Solidi B*, **244**(3), 910–918.

133. Lim TC (2012), Mixed auxeticity of auxetic sandwich structures, *Phys. Status Solidi B*, **249**(7), 1366–1372.

134. Lakes R (1987), Foam structures with a negative Poisson's ratio, *Science*, **235**(4792), 1038–1040.

135. Chan N and Evans KE (1998), Indentation resilience of conventional and auxetic foams, *J. Cell. Plast.*, **34**(3), 231–260.

136. Scarpa F, Yates JR, Ciffo LG, and Patsias S (2002), Dynamic crushing of auxetic open-cell polyurethane foam, *Proc. Inst. Mech. Eng. Part C: J. Mech. Eng. Sci.*, **216**(12), 1153–1156.

137. Friis EA, Lakes RS, and Park JB (1988), Negative Poisson's ratio polymeric and metallic foams, *J. Mater. Sci.*, **23**(12), 4406–4414.

138. Chen CP and Lakes RS (1996), Micromechanical analysis of dynamic behavior of conventional and negative Poisson's ratio foams, *J. Eng. Mater. Technol.*, **118**(3), 285–288.

139. Chan N and Evans KE (1999), The mechanical properties of conventional and auxetic foams. Part I: Compression and tension, *J. Cell. Plast.*, **35**(2), 130–165.

140. Chan N and Evans KE (1999), The mechanical properties of conventional and auxetic foams. Part II: Shear, *J. Cell. Plast.*, **35**(2), 166–183.

141. Bezazi A and Scarpa F (2007), Mechanical behaviour of conventional and negative Poisson's ratio thermoplastic polyurethane foams under compressive cyclic loading, *Int. J. Fatigue*, **29**(5), 922–930.

142. Bezazi A and Scarpa F (2009), Tensile fatigue of conventional and negative Poisson's ratio open cell PU foams, *Int. J. Fatigue*, **31**(3), 488–494.

143. Bezazi A, Boukharouba W, and Scarpa F (2009), Mechanical properties of auxetic carbon/epoxy composites: Static and cyclic fatigue behaviour, *Phys. Status Solidi B*, **246**(9), 2102–2110.

144. Lim TC, Alderson A, and Alderson KL (2014), Experimental studies on the impact properties of auxetic materials, *Phys. Status Solidi B*, **251**(2), 307–313.

145. Scarpa F, Giacomin J, Zhang Y, and Pastorino P (2005), Mechanical performance of auxetic polyurethane foam for antivibration glove applications, *Cell. Polym.*, **24**(5), 253–268.

146. Choi JB and Lakes R (1995), Analysis of elastic modulus of conventional foams and of re-entrant foam materials with a negative Poisson's ratio, *Int. J. Mech. Sci.*, **37**(1), 51–59.

147. Scarpa F, Giacomin JA, Bezazi A, and Bullough WA (2006), Dynamic behavior and damping capacity of auxetic foam pads, *Smart Structures and Materials 2006: Damping and Isolation*, *Proc. SPIE*, **6169**, 61690T.

148. Pastorino P, Scarpa F, Patsias S, Yates JR, Haake SJ, and Ruzzene M (2007), Strain rate dependence of stiffness and Poisson's ratio of auxetic open cell PU foams, *Phys. Status Solidi B*, **244**(3), 955–965.

149. Lowe A and Lakes RS (2000), Negative Poisson's ratio foam as seat cushion material, *Cell. Polym.*, **19**(3), 157–167.

150. Lim TC (2018), Auxeticity of concentric auxetic-conventional foam rods with high modulus interface adhesive, *Materials*, **11**(2), 223.

151. Strek AM (2010), Production and study of polyether auxetic foam, *Mech. Control*, **29**(2), 78–87.

152. Bianchi M, Scarpa F, Banse M, and Smith CW (2011), Novel generation of auxetic open cell foams for curved and arbitrary shapes, *Acta Mater.*, **59** (2), 686–691.

153. Bianchi M, Frontoni S, Scarpa F, and Smith CW (2011), Density change during the manufacturing process of PU-PE open cell auxetic foams, *Phys. Status Solidi B*, **248**(1), 30–38.

154. Alderson K, Alderson A, Ravirala N, Simkins V, and Davies P (2012), Manufacture and characterisation of thin flat and curved auxetic foam sheets, *Phys. Status Solidi B*, **249**(7), 1315–1321.

155. Alderson KL and Evans KE (1993), Strain-dependent behaviour of microporous polyethylene with a negative Poisson's ratio, *J. Mater. Sci.*, **28**(15), 4092–4098.

156. Alderson KL, Alderson A, Webber RS, and Evans KE (1998), Evidence for uniaxial drawing in the fibrillated microstructure of auxetic microporous polymers, *J. Mater. Sci. Lett.*, **17**(16), 1415–1419.

157. Pickles AP, Webber RS, Alderson KL, Neale PJ, and Evans KE (1995), The effect of the processing parameters on the fabrication of auxetic polyethylene, *J. Mater. Sci.*, **30**(16), 4059–4068.

158. Alderson KL, Webber RS, and Evans KE (2007), Microstructural evolution in the processing of auxetic microporous polymers, *Phys. Status Solidi B*, **244**(3), 828–841.

159. Ravirala N, Alderson A, Alderson KL, and Davies PJ (2005), Auxetic polypropylene films, *Polym. Eng. Sci.*, **45**(4), 517–528.

160. Ravirala N, Alderson KL, Davies PJ, Simkins VR, and Alderson A (2006), Negative Poisson's ratio polyester fibers, *Text. Res. J.*, **76**(7), 540–546.

161. Chirima GT, Zied KM, Ravirala N, Alderson KL, and Alderson A (2009), Numerical and analytical modelling of multi-layer adhesive-film interface systems, *Phys. Status Solidi B*, **246**(9), 2072–2082.

162. Evans KE (1991), Auxetic polymers: A new range of materials, *Endeavour*, **15**(4), 170–174.

163. Alderson KL, Pickles AP, Neale PJ, and Evans KE (1994), Auxetic polyethylene: The effect of a negative Poisson's ratio on hardness, *Acta Metall. Mater.*, **42**(7), 2261–2266.

164. Franke M and Magerle R (2011), Locally auxetic behavior of elastomeric polypropylene on the 100 nm length scale, *ACS Nano*, **5**(6), 4886–4891.

165. Baker J, Douglass A, and Griffin A (1995), Trimeric liquid crystals: Model compounds for auxetic polymer, *ACS-Polym. Prepr.*, **36**(2), 345–346.

166. Liu P, He C, and Griffin A (1998), Liquid crystalline polymers as potential auxetic materials: Influence of transverse rods on the polymer mesophase, *Abstr. Pap. Am. Chem. Soc.*, **216**, 108.

167. He C, Liu P, and Griffin A (1998), Toward negative Poisson ratio polymers through molecular design, *Macromol.*, **31**(9), 3145–3147, 1998.

168. He C, Liu P, Griffin A, Smith CW, and Evans KE (2005), Morphology and deformation behaviour of a liquid crystalline polymer containing laterally attached pentaphenyl rods, *Macromol. Chem. Phys.*, **206**(2), 233–239.

169. He C, Liu P, McMullan PJ, and Griffin AC (2005), Toward molecular auxetics: Main chain liquid crystalline polymers consisting of laterally attached para-quaterphenyls, *Phys. Status Solidi B*, **242**, 576–584.

170. Li C, Xie X, and Cao S (2002), Synthesis and characterization of liquid crystalline copolyesters containing horizontal and lateral rods in main chain, *Polym. Adv. Technol.*, **13**(3–4), 178–187.

171. Dey S, Agra-Kooijman DM, Ren W, McMullan PJ, Griffin AC, and Kumar S (2013), Soft elasticity in main chain liquid crystal elastomers, *Crystals*, **3**(2), 363–390.

172. Baughman RH and Galvao DS (1993), Crystalline networks with unusual predicted mechanical and thermal properties, *Nature*, **365**(6448), 735–737.

173. Baughman RH, Galvao DS, Cui C, and Dantas S (1997), Hinged and chiral polydiacetylene carbon crystals, *Chem. Phys. Lett.*, **269**(3–4), 356–364.

174. Nkansah MA, Evans KE, and Hutchinson IJ (1994), Modelling the mechanical properties of an auxetic molecular network, *Model. Simul. Mater. Sci. Eng.*, **2**(3), 337–352.

175. Grima JN and Evans KE (2000), Self-expanding molecular networks, *Chem. Comm.*, **2000**(16), 1531–1532.

176. Grima JN (2008), On the mechanical properties and auxetic potential of various organic networked polymers, *Mol. Simul.*, **34**(10–15), 1149–1158.

177. Wen X, Garland CW, Hwa T, Kardar M, Kokufuta E, Li Y, Orkisz M, and Tanaka T (1992), Crumpled and collapsed conformation in graphite oxide membranes, *Nature*, **355**(6359), 426–428.

178. Pour N, Itzhaki L, Hoz B, Altus E, Basch H, and Hoz S (2006), Auxetics at the molecular level: A negative Poisson's ratio in molecular rods, *Angew. Chemie*, **45**(36), 5981–5983.

179. Pour N, Altus E, Basch H, and Hoz S (2009), The origin of the auxetic effect in prismanes: Bowtie structure and the mechanical properties of biprismanes, *J. Phys. Chem. C*, **113**(9), 3467–3470.

180. Hall LJ, Coluci VR, Galvao DS, Kozlov ME, Dantas SO, and Baughman RH (2008), Sign change of Poisson's ratio for carbon nanotube sheets, *Science*, **320**(5875), 504–507.

181. Yao YT, Alderson A, and Alderson KL (2008), Can nanotubes display auxetic behaviour?, *Phys. Status Solidi B*, **245**(11), 2373–2382.

182. Chen L, Liu C, Wang J, Zhang W, Hu C, and Fan S (2009), Auxetic materials with large negative Poisson's ratios based on highly oriented carbon nanotube structures, *Appl. Phys. Lett.*, **94**(25), 253111.

183. Scarpa F, Adhikari S, and Srikantha Phani A (2009), Effective elastic mechanical properties of single layer graphene sheets, *Nanotechnol.*, **20**(6), 065709.

184. Braghin FL and Hasselmann N (2010), Thermal fluctuations of free-standing graphene, *Phys. Rev. B*, **82**(3), 035407.

185. Cadelano E, Palla PL, Giordano S, and Colombo L (2010), Elastic properties of hydrogenated graphene, *Phys. Rev. B*, **82**(23), 235414.

186. Sihn S, Varshney V, Roy AK, and Farmer BL (2012), Prediction of 3D elastic moduli and Poisson's ratios of pillared graphene nanostructures, *Carbon*, **50**(2), 603–611.

187. Koskinen P (2014), Graphene cardboard: From ripples to tunable metamaterial, *Appl. Phys. Lett.*, **104**(10), 101902.

188. Grima JN, Jackson R, Alderson A, and Evans KE (2000), Do zeolites have negative Poisson's ratios?, *Adv. Mater.*, **12**(24), 1912–1918.

189. Grima JN (2000), *New auxetic materials*, Ph.D Thesis, University of Exeter, Exeter, UK.

190. Sanchez-Valle C, Sinogeikin SV, Lethbridge ZAD, Walton RI, Smith CW, Evans KE, and Bass JD (2005), Brillouin scattering study on the single-crystal elastic properties of natrolite and analcime zeolites, *J. Appl. Phys.*, **98**(5), 053508.

191. Lethbridge ZAD, Williams JJ, Walton RI, Smith CW, Hooper RM, and Evans KE (2006), Direct, static measurement of single-crystal Young's moduli of the zeolite natrolite: Comparison with dynamic studies and simulations, *Acta Mater.*, **54**(9), 2533–2545.

192. Smith CW, Evans KE, Lethbridge ZAD, and Walton RI (2007), An analytical model for producing negative Poisson's ratios and its application in explaining off-axis elastic properties of the NAT-type zeolites, *Acta Mater.*, **55**(17), 5697–5707.

193. Williams JJ, Smith CW, Evans KE, Lethbridge ZAD, and Walton RI (2007), Off-axis elastic properties and the effect of extraframework species on structural flexibility of the NAT-type zeolites: Simulations of structure and elastic properties, *Chem. Mater.*, **19**(10), 2423–2434.

194. Grima JN, Gatt R, Zammit V, Williams JJ, Alderson A, and Walton RI (2007), Natrolite: A zeolite with negative Poisson's ratios, *J. Appl. Phys.*, **101**(8), 086102.

195. Grima JN, Zammit V, Gatt R, Alderson A, and Evans KE (2007), Auxetic behaviour from rotating semi-rigid units, *Phys. Status Solidi B*, **224**(3), 866–882.

196. Grima JN, Gatt R, and Chetcuti E (2010), On the behaviour of natrolite under hydrostatic pressure, *J. Non-Cryst. Solids*, **356**(37–40), 1881–1887.

197. Kremleva A, Vogt T, and Rosch N (2013), Monovalent cation exchanged natrolites and their behaviour under pressure. A computational study, *J. Phys. Chem. C*, **117**(37), 19020–19030.

198. Manga ED, Blasco H, Da-Costa P, Drobek M, Ayral A, Le Clezio E, Despaux G, Coasne B, and Julbe A (2014), Effect of gas adsorption on acoustic wave propagation in MFI zeolite membrane materials: Experiment and molecular simulation, *Langmuir*, **30**(34), 10336–10343.

199. Siddorn M, Coudert FX, Evans KE, and Marmier AA (2015), A systematic typology for negative Poisson's ratio materials and the prediction of complete auxeticity in pure silica zeolite JST, *Phys. Chem. Chem. Phys.*, **17**(27), 17927.

200. Alderson A and Evans KE (2002), Molecular origin of auxetic behaviour in tetrahedral framework silicates, *Phys. Rev. Lett.*, **89**(22), 225503.

201. Alderson A, Alderson KL, Evans KE, Grima JN, Williams MR, and Davies PJ (2004), Molecular modelling of the deformation mechanisms acting in auxetic silica, *Comput. Meth. Sci. Technol.*, **10**(2), 117–126.

202. Alderson A, Alderson KL, Evans KE, Grima JN, and Williams M (2005), Modelling of negative Poisson's ratio nanomaterials: Deformation mechanisms, structure-property relationships and applications, *J. Metastab. Nanocryst. Mater.*, **23**, 55–58.

203. Gatt R, Mizzi L, Azzopardi KM, and Grima JN (2015), A force-field based analysis of the deformation in α-cristobalite, *Phys. Status Solidi B*, **252**(7), 1479–1485.

204. Nazare F and Alderson A (2015), Models for the prediction of Poisson's ratio in the α-cristobalite tetrahedral framework, *Phys. Status Solidi B*, **252**(7), pp. 1465–1478.

205. Wojciechowski KW (1989), Two-dimensional isotropic system with a negative Poisson ratio, *Phys. Lett. A*, **137**(1–2), 60–64.

206. Wojciechowski KW, Branka AC, and Parrinello M (1984), Monte Carlo study of the phase diagram of a two dimensional system of hard cyclic hexamers, *Mol. Phys.*, **53**(6), 1541–1545.

207. Wojciechowski KW (1987), Constant thermodynamic tension Monte Carlo studies of elastic properties of a two-dimensional system of hard cyclic hexamers, *Mol. Phys.*, **61**(5), 1247–1258.

208. Wojciechowski KW and Branka AC (1989), Negative Poisson ratio in a two-dimensional "isotropic" solid, *Phys. Lett. A*, **40**(12), 7222–7225.

209. Wojciechowski KW and Branka AC (1994), Auxetics: Materials and models with negative Poisson's ratios, *Mol. Phys. Rep.*, **6**, 71.

210. Tretiakov KV and Wojciechowski KW (2002), Orientational phase transition between hexagonal solids in planar systems of hard cyclic pentamers and heptamers, *J. Phys.: Condens. Matter*, **14**(6), 1261–1273.

211. Tretiakov KV and Wojciechowski KW (2005), Monte Carlo simulation of two-dimensional hard body systems with extreme values of the Poisson's ratio, *Phys. Status Solidi B*, **242**(3), 730–741.

212. Tretiakov KV and Wojciechowski KW (2006), Elastic properties of the degenerate crystalline phase of two-dimensional hard dimers, *J. Non-Cryst. Solids*, **352**(40–41), 4221–4228.

213. Tretiakov KV and Wojciechowski KW (2007), Poisson's ratio of simple planar "isotropic" solids in two dimensions, *Phys. Status Solidi B*, **244**(3), 1038–1046.

214. Narojczyk JW and Wojciechowski KW (2007), Elastic properties of two-dimensional soft polydisperse trimers at zero temperature, *Phys. Status Solidi B*, **244**(3), 943–954.

215. Narojczyk JW, Alderson A, Imre AR, Scarpa F, and Wojciechowski KW (2008), Negative Poisson's ratio behaviour in the planar model of asymmetric trimers at zero temperature, *J. Non-Cryst. Solids*, **354**(35–39), 4242–4248.

216. Narojczyk JW and Wojciechowski KW (2008), Elasticity of periodic and aperiodic structures of polydisperse dimers in two dimensions at zero temperature, *Phys. Status Solidi B*, **245**(11), 2463–2468.

217. Narojczyk JW and Wojciechowski KW (2010), Elastic properties of degenerate f.c.c. crystal of polydisperse soft dimers at zero temperature, *J. Non-Cryst. Solids*, **356**(37–40), 2026–2032.

218. Narojczyk JW, Piglowski PM, Wojciechowski KW, and Tretiakov, KV (2015), Elastic properties of mono and polydisperse two dimensional crystals of hard core repulsive Yukawa particles, *Phys. Status Solidi B*, **252**(7), 1508–1513.

219. Tretiakov KV, Piglowski PM, Hyzorek K, and Wojciechowski KW (2016), Enhanced auxeticity in Yukawa systems due to introduction of nanochannels in-direction, *Smart Mater. Struct.*, **25**(5), 054007.

220. Lakes R (1993), Advances in negative Poisson's ratio materials, *Adv. Mater.*, **5**(4), 293–296.

221. Alderson A (1999), A triumph of lateral thought, *Chemistry & Industry*, **1993**(10), 384–391.

222. Yang W, Li ZM, Shi W, Xie BH, and Yang MB (2004), Review on auxetic materials, *J. Mater. Sci.*, **39**(10), 3269–3279.

223. Alderson A and Alderson KL (2007), Auxetic materials, *IMechE J. Aerosp. Eng.*, **221**(4), 565–575.

224. Liu Y and Hu H (2010), A review on auxetic structures and polymeric materials, *Scient. Res. Essays*, **5**(10), 1052–1063.

225. Greaves GN, Greer AL, Lakes RS, and Rouxel T (2011), Poisson's ratio and modern materials, *Nat. Mater.*, **10**(11), 823–837.

226. Prawoto Y (2012), Seeing auxetic materials from the mechanics point of view: A structural review on the negative Poisson's ratio, *Comput. Mater. Sci.*, **58**, 140–153.

227. Critchley R, Corni I, Wharton JA, Walsh FC, Wood RJK, and Stokes KR (2013), A review of the manufacture, mechanical properties and potential applications of auxetic foams, *Phys. Status Solidi B*, **250**(10), 1963–1982.

228. Saxena KK, Das R, and Calius EP (2016), Three decades of auxetics research – materials with negative Poisson's ratio: A review, *Adv. Eng. Mater.*, **18**(11), 1847–1870.

229. Lakes RS (2017), Negative-Poisson's-ratio materials: Auxetic solids, *Ann. Rev. Mater. Res.*, **47**, 63–81.

230. Lim TC (2017), Analogies across auxetic models based on deformation mechanism, *Phys. Status Solidi RRL*, **11**(6), 1600440.

231. Kolken HMA and Zadpoor AA (2017), Auxetic mechanical metamaterials, *RSC Adv.*, **7**(9), 5111–5129.

232. Ren X, Das R, Tran P, Ngo T, and Xie YM (2018), Auxetic metamaterials and structures: A review, *Smart Mater. Struct.*, **27**(2), 023001.

233. Darja R, Tatjana R, and Alenka PC (2013), Auxetic textiles, *Acta Chim. Slov.*, **60**(4), 715–723.

234. Jiang JW, Kim SY, and Park HS (2016), Auxetic nanomaterials: Recent progress and future development, *Appl. Phys. Rev.*, **3**, 041101.

235. Park HS and Kim SY (2017), A perspective on auxetic nanomaterials, *Nano Converg.*, **4**, 10.

236. Duncan O, Shepherd T, Moroney C, Foster L, Venkatraman PD, Winwood K, Allen T, and Alderson A (2018), Review of auxetic materials for sports applications: Expanding options in comfort and protection, *Appl. Sci.*, **8**(6), 941.

237. Kwietniewski M and Miedzińska D (2019), Review of elastomeric materials for application to composites reinforced by auxetics fabrics, *Proc. Struct. Integrity*, **17**, 154–161.

238. Wu W, Hu W, Qian G, Liao H, Xu X, and Berto F (2019), Mechanical design and multifunctional applications of chiral mechanical metamaterials: A review, *Mater. Des.*, **180**, 107950.

239. Zhang J, Lu G, and You Z (2020), Large deformation and energy absorption of additively manufactured auxetic materials and structure: A review, *Compos. Part B: Eng.*, **201**, 108340.

240. Lim TC (2015), *Auxetic Materials and Structures*, Springer Nature, Singapore.

241. Hu H, Zhang M, and Liu Y (2019), *Auxetic Textiles*, Woodhead, Duxford.

242. Lim TC (2020), *Mechanics of Metamaterials with Negative Parameters*, Springer Nature.

243. Grether GF, Kolluru GR, and Nersissian K (2004), Individual colour patches as multicomponent signals, *Biol. Rev. (Camb. Philos. Soc.)*, **79**(3), 583–610.

244. Saenko SV, Teyssier J, van der Marel D, and Milinkovitch MC (2013), Precise colocalization of interacting structural and pigmentary elements generates extensive color pattern variation in Phelsuma lizards, *BMC Biol.*, **11**, 105.

245. Teyssier J, Saenko SV, van der Marel, and Milinkovitch MC (2015), Photonic crystals cause active colour change in chameleons, *Nat. Commun.*, **6**, 6368.

246. Hedayati MK and Elbahri M (2017), Review of metasurface plasmonic structural color, *Plasmonics*, **12**(5), 1463–1479.

247. Galinski H, Favraud G, Dong H, Gongora JST, Favaro G, Döbeli M, Spolenak R, Fratalocchi A, and Capasso F (2017), Scalable, ultra-resistant structural colors based on network metamaterials, *Light: Sci. Appl.*, **6**, e16233.

248. Huang L, Duan Y, Dai X, Zeng Y, Ma G, Liu Y, Gao S, and Zhang W (2019), Bioinspired metamaterials: Multibands electromagnetic wave adaptability and hydrophobic characteristics, *Small* **15**(40), 1902730.

249. Lee N, Kim T, Lim JS, Chang I, and Cho HH (2019), Metamaterial-selective emitter for maximizing infrared camouflage performance with energy dissipation, *Appl. Mater. Interf.*, **11**(23), 21250–21257.

250. Rafsanjani A and Pasini D (2016), Multistable compliant auxetic meta-materials inspired by geometric patterns in Islamic arts, *Bull. Am. Phys. Soc.*, **61**, K40.00008.

251. Rafsanjani A and Pasini D (2016), Bistable auxetic mechanical metama-terials inspired by ancient geometric motifs, *Extreme Mech. Lett.*, **9**, 291–296.

252. Ravirala N, Alderson A, and Alderson KL (2007), Interlocking hexagonal model for auxetic behaviour, *J. Mater. Sci.*, **42**(17), 7433–7445, 2007.

253. Lim TC (2021), A perfect 2D auxetic sliding mechanism based on an Islamic geometric pattern, *Eng. Res. Express*, **3**(1), 015025.

254. Larsen UD, Sigmund O, and Bouwstra S (1997), Design and fabrication of compliant mechanisms and material structures with negative Poisson's ratio, *J. Microelectromech. Syst.*, **6**(2), 99–106.

255. Li Y, Chen Y, Li T, Cao S, and Wang L (2018), Hoberman-sphere-inspired lattice metamaterials with tunable negative thermal expansion, *Compos. Struct.*, **189**, 586–597.

256. Lim TC (2020), Metacomposite structure with sign-changing coefficients of hygrothermal expansions inspired by Islamic motif, *Compos. Struct.*, **251**, 112660.

257. Cabras L, Brun M, and Misseroni D (2019), Micro-structured medium with large isotropic negative thermal expansion, *Proc. Royal Soc A*, **475** (2232), 20190463.

258. Lim TC (2021), An auxetic system based on interconnected Y-elements inspired by Islamic geometric patterns, *Symmetry*, **13**(5), 865.

259. Lim TC (2021), Aspect ratio and size effects of a metacomposite with interconnected Y-elements, *J. Phys.: Conf. Ser.*, **2047**, 012029.

260. Lakes R (1996), Cellular solid structures with unbounded thermal expan-sion, *J. Mater. Sci. Lett.*, **15**(6), 475–477.

261. Lakes R (2007), Cellular solids with tunable positive or negative thermal expansion of unbounded magnitude, *Appl. Phys. Lett.*, **90**(22), 221905.

262. Lim TC (2020), Composite metamaterial square grids with sign-flipping expansion coefficients leading to a type of Islamic design, *SN Appl. Sci.*, **2**(5), 918.

263. Lim TC (2021), Metamaterial honeycomb with sign-toggling expansion coefficients that manifests an Islamic mosaic pattern at the Alhambra Palace, *Adv. Compos. Hybrid Mater.*, **4**(4), 966–978.

264. Lim TC (2022), Metamaterial with sign-toggling thermal expansivity inspired by Islamic motifs in Spain, *J. Sci.: Adv. Mater. Dev.*, **7**(1), 100401.

265. Hashemi MRM, Yang SH, Wang T, Sepúlveda N, and Jarrahi M (2016), Electronically-controlled beam-steering through vanadium dioxide metasurfaces, *Scient. Rep.* **6**, 35439.

266. Wang GD, Liu MH, Hu XW, Kong LH, Cheng LL, and Chen ZQ (2013), Multi-band microwave metamaterial absorber based on coplanar Jerusalem crosses, *Chin. Phys. B*, **23**(1), 017802.

267. Arezoomand AS, Zarrabi FB, Heydari S, and Gandji NP (2015), Independent polarization and multi-band THz absorber base on Jerusalem cross, *Optics Commun.*, **352**, 121–126.

268. Mi S and Lee HY (2016), Design of a compact patch antenna loading periodic Jerusalem crosses, *Prog. Electromag. Res. M*, **47**, 151–159.

269. Naser-Moghadasi M, Nia AZ, Toolabi M, and Heydari S (2017), Microwave metamaterial absorber based on Jerusalem cross with meandered load for bandwidth enhancement, *Optik*, **140**, 515–522.

270. Silva Filho HVH, Silva CPN, de Oliveira MRT, de Oliveira EMF, de Melo MT, de Sousa TR, and Gomes Neto A (2017), Multiband FSS with fractal characteristic based on Jerusalem cross geometry, *J. Microwaves Optoelectron. Electromag. Appl.*, **16**(4), 932–941.

271. Jafari FS, Naderi M, Hatami A, and Zarrabi FB (2019), Microwave Jerusalem cross absorber by metamaterial split ring resonator load to obtain polarization independence with triple band application, *AEU – Int. J. Electron. Commun.*, **101**(5), 138–144.

272. Kamonsin W, Krachodnok P, Chomtong P, and Akkaraekthalin P (2020), Dual-band metamaterial based on Jerusalem cross structure with inter-digital technique for LTE and WLAN systems, *IEEE Access*, **8**, 21565–21572.

273. Tang Y, He L, Liu A, Xiong C, and Xu H (2020), Optically transparent metamaterial absorber based on Jerusalem cross structure at S-band frequencies, *Modern Phys. Lett. B*, **34**(16), 2050175.

274. Hannan S, Islam MT, Almutairi AF, and Faruque MRI (2020), Wide bandwidth angle- and polarization-insensitive symmetric metamaterial absorber for X and Ku band applications, *Scient. Rep.*, **10**, 10338.

275. Liu Q, Liang D, Wang X, Han T, Lu H, and Xie J (2021), Jerusalem cross geometry magnetic substrate absorbers for low-frequency broadband applications, *AIP Adv.*, **11**(3), 035037.

276. Lee SC, Kang JH, Park QH, Krishna S, and Brueck SRJ (2016), Oscillatory penetration of near-fields in plasmonic excitation at metal-dielectric interfaces, *Scient. Rep.*, **6**, 24400.

277. Lim TC (2019), An anisotropic auxetic 2D metamaterial based on sliding microstructural mechanism, *Mater.*, **12**(3), 429.

278. Cabras L and Brun M (2014), Auxetic two-dimensional lattices with Poisson's ratio arbitrarily close to −1, *Proc. Royal. Soc. A*, **470**(2172), 20140538.

279. Lim TC (2021), An auxetic metamaterial with tunable positive to negative hygrothermal expansion by means of counter-rotating crosses, *Phys. Status Solidi B*, **258**(8), 2100137.

280. Zhu WM, Liu AQ, Bourouina T, Tsai DP, Teng JH, Zhang XH, Lo GQ, Kwong DL, and Zheludev NI (2012), Microelectromechanical Maltese-cross metamaterial with tunable terahertz anisotropy, *Nat. Commun.*, **3**, 1274.

281. Lim TC (2021), Adjustable positive and negative hygrothermal expansion metamaterial inspired by the Maltese cross, *Royal Soc. Open Sci.*, **8**(8), 210593.

282. Evans KE and Alderson A (2000), Auxetic materials: Functional materials and structures from lateral thinking!, *Adv. Mater.*, **12**(9), 617–628.

283. Cheng HC, Scarpa F, Panzera TH, Farrow I, and Peng HX (2019), Shear stiffness and energy absorption of auxetic open cell foams as sandwich cores, *Phys. Status Solidi B*, **256**(1), 1800411.

284. Bliven E, Rouhier A, Tsai S, Wilinger R, Bourdet N, Deck C, Madey SM, and Bottlang M (2019), Evaluation of a novel bicycle helmet concept in oblique impact testing, *Accident Anal. Prevent.*, **124**, 58–65.

285. Smith CW, Grima JN, and Evans KE (2000), A novel mechanism for generating auxetic behaviour in reticulated foams: Missing rib foam model, *Acta Mater.*, **48**(17), 4349–4356.

286. Zhong R, Fu M, Yin Q, Xu O, and Hu L (2019), Special characteristics of tetrachiral honeycombs under large deformation, *Int. J. Solids Struct.*, **169**, 166–176.

287. Alderson A, Alderson KL, Attard D, Evans KE, Gatt R, Grima JN, Miler W, Ravirala N, Smith CW, and Zied K (2010), Elastic constants of 3-, 4- and

6-connected chiral and anti-chiral honeycombs subject to uniaxial in-plane loading, *Compos. Sci. Technol.*, **70**(7), 1042–1048.

288. Caruana-Gauci R, Degabriele EP, Attard D, and Grima JN (2018), Auxetic metamaterial inspired from wine-racks, *J. Mater. Sci.*, **53**(7), 5079–5091.

289. Gatt R, Mizzi L, Azzopardi JI, Azzopardi KM, Attard D, Casha A, Briffa J, and Grima JN (2015), Hierarchical auxetic mechanical metamaterials, *Scient. Rep.*, **5**, 8395.

290. Meza LR, Zelhofer AJ, Clarke N, Mateos AJ, Kochmann DM, and Greer JR (2015), Resilient 3D hierarchical architected metamaterials, *Proc. Nat. Acad. Sci.*, **112**(37), 11502–11507.

291. Mousanezhad D, Babaee S, Ebrahimi H, Ghosh R, Hamouda AS, Bertoldi K, and Vaziri A (2016), Hierarchical honeycomb auxetic metamaterials, *Scient. Rep.*, **5**, 18306.

292. Chen Y, Jia Z, and Wang L (2016), Hierarchical honeycomb lattice metamaterials with improved thermal resistance and mechanical properties, *Compos. Struct.*, **152**, 395–402.

293. Grima-Cornish JN, Grima JN, and Evans KE (2017), On the structural and mechanical properties of poly(phenylacetylene) truss-like hexagonal hierarchical nanonetworks, *Phys. Status Solidi B*, **254**(12), 1700190.

294. Billon K, Zampetakis I, Scarpa F, Ouisse M, Sadoulet-Reboul E, Collet M, Perriman A, and Hetherington A (2017), Mechanics and band gaps in hierarchical auxetic rectangular perforated composite metamaterials, *Compos. Struct.*, **160**, 1042–1050.

295. Yang H, Wang B, and Ma L (2019), Designing hierarchical metamaterials by topology analysis with tailored Poisson's ratio and Young's modulus, *Compos. Struct.*, **214**, 359–378.

296. Zhang W, Zhao S, Sun R, Scarpa F, and Wang J (2019), In-plane mechanical behavior of a new star-re-entrant hierarchical metamaterial, *Polym.*, **11**(7), 1132.

297. Kochmann DM, Hopkins JB, and Valdevit L (2019), Multiscale modeling and optimization of the mechanics of hierarchical metamaterials, *MRS Bull.*, **44**(10), 773–781.

298. Zhang H, Wu J, Fang D, and Zhang Y (2021), Hierarchical mechanical metamaterials built with scalable tristable elements for ternary logic operation and amplitude modulation, *Sci. Adv.*, **7**(9), eabf1966.

299. Guo X, Ni X, Li J, Zhang H, Zhang F, Yu H, Wu J, Bai Y, Lei H, Huang Y, Rogers JA, and Zhang H (2021), Designing mechanical metamaterials with kirigami inspired, hierarchical constructions for giant positive and negative thermal expansion, *Adv. Mater.*, **33**(3), 2004919.

300. Dudek KK, Gatt R, Dudel MR, and Grima JN (2021), Controllable hierarchical mechanical metamaterials guided by the hinge design, *Materials*, **14**(4), 758.

301. Morvaridi M, Carta G, Bosia F, Gliozzi AS, Pugno NM, Misseroni D, and Brun M (2021), Hierarchical auxetic and isotropic porous medium with extremely negative Poisson's ratio, *Extreme Mech. Lett.*, **48**, 101405.

302. Wu H, Cheng Y, Liu W, He R, Zhou M, Wu S, Song X, and Chen Y (2016), Effect of the particle size and the debinding process on the density of alumina ceramics fabricated by 3D printing based on stereolithography, *Ceram. Int.*, **42**(15), 17290–17294.

303. Domènech B, Tan ATL, Jelitto H, Berodt EZ, Blankenburg M, Focke O, Cann J, Tasan CC, Ciacchi LC, Müller M, Furlan KP, Hart AJ, and Schneider GA (2020), Strong macroscale supercrystalline structures by 3D printing combined with self-assembly of ceramic functionalized nanoparticles, *Adv. Eng. Mater.*, **22** (7), 2000352.

Cambridge Elements ☰

Emerging Theories and Technologies in Metamaterials

Tie Jun Cui

Southeast University, China

Tie Jun Cui is Cheung-Kong Professor and Chief Professor at Southeast University, China, and a Fellow of the IEEE. He has made significant contributions to the area of effective-medium metamaterials and spoof surface plasmon polaritons at microwave frequencies, both in new-physics verification and engineering applications. He has recently proposed digital coding, field-programmable, and information metamaterials, which extend the concept of metamaterial.

John B. Pendry

Imperial College London

Sir John Pendry is Chair in Theoretical Solid State Physics at Imperial College London, and a Fellow of the Royal Society, the Institute of Physics and the Optical Society of America. Among his many achievements are the proposal of the concepts of an 'invisibility cloak' and the invention of the transformation optics technique for the control of electromagnetic fields.

About the Series

This series systematically covers the theory, characterisation, design and fabrication of metamaterials in areas such as electromagnetics and optics, plasmonics, acoustics and thermal science, nanoelectronics, and nanophotonics, and also showcases the very latest experimental techniques and applications. Presenting cutting-edge research and novel results in a timely, indepth and yet digestible way, it is perfect for graduate students, researchers, and professionals working on metamaterials.

Cambridge Elements ≡

Emerging Theories and Technologies in Metamaterials

Elements in the Series

Spoof Surface Plasmon Metamaterials
Paloma Arroyo Huidobro, Antonio I. Fernández-Domíguez, John B. Pendry,
Luis Martín-Moreno, Francisco J. Garcia-Vidal

Metamaterials and Negative Refraction
Rujiang Li, Zuojia Wang, and Hongsheng Chen

Information Metamaterials
Tie Jun Cui and Shuo Liu

Effective Medium Theory of Metamaterials and Metasurfaces
Wei Xiang Jiang, Zhong Lei Mei, and Tie Jun Cui

Epsilon-Near-Zero Metamaterials
Yue Li, Ziheng Zhou, Yijing He, and Hao Li

A Partially Auxetic Metamaterial Inspired by the Maltese Cross
Teik-Cheng Lim

A full series listing is available at: www.cambridge.org/EMMA

Printed in the United States
by Baker & Taylor Publisher Services

Printed in the United States
by Baker & Taylor Publisher Services